Professor Bear's Math Club
Workbook Alpha

Marjorie Sayer

Atelier Finwhale

Preface

This book arose from my work as a volunteer math coach for an elementary school math club.

The emphasis of this book is **not** the elementary school math curriculum;
not "bringing kids up to international standards;"
not "something relevant to the 21st century."

The emphasis **is**: engaging in mathematics as a life-affirming creative pursuit.

This is a short way of saying that mathematics is fun; it connects you to other people who are interested in having fun; it gives you opportunities to use the math parts of your brain—and everyone has them.

When you do fun math, when you connect with other people doing fun math, and when you use your math abilities to your full potential, mathematics makes you feel glad to be you.

It is my great pleasure to write this book with no other ideal in mind.

RRRRR.

Professor L. F. Bear
Palo Alto, 2012

Acknowledgements

Professor L. F. Bear gratefully acknowledges the inspiration provided to her by the members of the El Carmelo Math Club: Arjun, Ava, Chris, Ciel, Clara, Harrison, Jeffrey, Jonathon, Kristina, Lauren, Lia, Miranda, Moeka, Noam, Sarah, Sophia.

For encouragement and valuable insights, Professor Bear owes much to Jeannette Cheng, Gabrielle Conway, Betsy Franco, Alan Liu, Heidi Mickelsen, Ken Tracton, Maya Venkatraman, and Stephanie Youngquist. Thank you so very much.

Any errors or omissions are due to Marjorie Sayer.

How To Use This Workbook

This workbook is designed to be your math laboratory, where you store your equipment and do your experiments. Typical cub age for this book is 8-11 years.

The workbook has ten chapters, on ten topics. Every chapter has this structure:

Title	Tells you what the topic is. Feel free to customize this page.
Notes and Terminology	This section is an introduction to the topic. You can work through this section with friends and maybe a coach. An important part of every introduction is vocabulary—it's essential for bears and humans to agree on the words they use with one another. This section might also describe some techniques for solving specific kinds of problems. Think of these sections as a dialogue between you and Professor Bear. Feel free to write down "What?" if you don't get something, or if you disagree with something. And then hunt for a more satisfying answer. Think about the topic, ask your colleagues, and as a last resort ask your coach. And then write down whatever makes the most sense to you. Professor Bear hopes that you will come back to these pages when you forget some little fact (all mathematicians forget something, sometime) and you will have a record of what you found out here. Above all, clarify these notes as needed. Customize them.
Problems 1 through 10	Every chapter has ten problems on that topic. The problems tend to get more challenging as the number increases. Some problems have specific answers such as 42. Other problems are for you to think about. There is a page or two of paper devoted to each problem. Use as much space as you need to find solutions.
How To Deal With <Topic> Problems	This is a page for you to write down any tips, patterns, gotchas, or Grand Unified Theories that you have discovered about the topic. A place for considered reflection.

Problem solutions are listed after the tenth chapter. More detailed solutions and discussion are in a separate book: *Professor Bear's Coach Manual Alpha*.
See also: http://bear.mathbelt.com

Contents

1. Sequences

1, 2, 3, . . .

duck, duck, goose, . . .

4, 3, 2, 1

Sequences: Notes and Vocabulary

A **sequence** is an ordered list of things.

Here is a sequence of numbers: 1, 2, 3, 4, 5
Here is a sequence of letters: A, B, C, D
Here is a different sequence of numbers: 5, 4, 3, 2, 1

The "things" in the sequence are called **terms**. In the sequence 5, 4, 3, 2, 1
- first term is 5
- the third term is 3
- the fifth term is 1.

What is the first term of the sequence 1, 2, 3, 4, 5? _____

What is the fourth term of the sequence A, B, C, D? _____

A sequence can be finite or infinite. The sequences above are all finite.

Here's another example of a finite sequence: 10, 9, 8

Here's another example: 400, 500, 600, 700, 800

Write down your own example of a finite sequence: _____

An infinite sequence is a list that goes on forever. It has an infinite number of terms. We show that it goes on forever by three periods in a row at the end: . . .

Another name for "three periods in a row" is ellipsis.

Here is an example of an infinite sequence: 1, 2, 3, . . .

Another example of an infinite sequence: $\dfrac{1}{2}, \dfrac{2}{3}, \dfrac{3}{4}$ . . .

Write down your own example of an infinite sequence: _____

A sequence of numbers might follow a pattern. If you know the pattern, you can figure out what the next term is. Can you write the next few terms of these sequences?

2, 4, 6, _____ , _____ , _____ , · · ·

4, 7, 10, _____ , _____ , _____ , · · ·

Some sequence patterns might be more challenging, such as this finite sequence:

3, 2, 1, 2, 3, 3, 3, 2, 2, 2, 3, 5, 5, ... , 1

Can you fill in the missing terms? Think for a minute. A hint for this sequence is at the bottom of this page.

You can also use an ellipsis to write down a missing set of terms in a finite sequence.
For example: 1, 2, 3, ... , 10
Is a short way to write 1, 2, 3, 4, 5, 6, 7, 8, 9, 10
In the example 1, 2, 3, ... , 10 the ellipsis stands for the missing terms 4, 5, 6, 7, 8, 9

In the following finite sequences, what are the missing terms?

Sequence	Missing Terms
10, 9, 8, ... , 4	
0, 3, 6, ... , 24	
77, 72, 67, ... , 47	

Hint for the sequence 3, 2, 1, 2, 3, 3, 3, 2, 2, 2, 3, 5, 5, ... , 1 above: musical notes. The sequence corresponds to a melody. If you don't have musical experience, this example will not mean much to you.

In that case, try to figure out the following pattern:

0, 3, 8, 15, 24, 35, 48, . . .

1-1.   Sample Sequences
Here are some sequences. What are the next three terms?

a.   5, 5.3, 5.6, 5.9, 6.2, . . .

b.   2, 6, 18, 54, 162, . . .

c.   2, 5, 9, 14, 20, . . .

d.   one, one thousand, one million, . . .

1-2.   Consecutive Terms

Consecutive terms are terms that exactly follow one another in a sequence.
In the sequence 1, 2, 3, ... the terms 3, 4, 5, 6, and 7 are consecutive terms.
In the sequence A, B, C, ... , Z the terms P and Q are consecutive terms.

Write three consecutive whole numbers: _____, _____ , _____

Write three consecutive whole numbers bigger than a million:

_____ , _____ , _____

Write three consecutive odd numbers: _____, _____ , _____

Write three consecutive multiples of five: _____ , _____ , _____

Write three consecutive prime numbers: _____ , _____ , _____

Write three consecutive two-digit primes: _____ , _____ , _____

Write three consecutive perfect squares: _____ , _____ , _____

Pose a more challenging problem involving consecutive numbers that you are pretty sure you are able to solve. Make it as different from the above problems as you can.

Example: Write three consecutive four-digit multiples of 17.

Pose an even more challenging problem involving consecutive numbers that you are pretty sure you are not able to solve (yet).

Example: What are the dates of the next three consecutive supernova explosions that will happen in the Milky Way?

1-3. Fibonacci Sequence

The Fibonacci sequence is: 1, 1, 2, 3, 5, 8, 13, . . .

a. The terms of the Fibonacci sequence follow a pattern. Describe the pattern.

b. List the next three terms of the Fibonacci sequence.

c. The terms of the Fibonacci sequence are called Fibonacci numbers. How many prime Fibonacci numbers can you find?

d. Choose any three consecutive terms of the Fibonacci sequence. Square the middle term and multiply the two other terms. Compare the results. Repeat this process with other groups of three consecutive terms. What is the pattern? (This is Cassini's Identity.)

1-4.   Zeno's Paradox In A Box

The sequence $\frac{1}{2}, \frac{1}{4}, \frac{1}{8}, \ldots$

Is illustrated in this box:

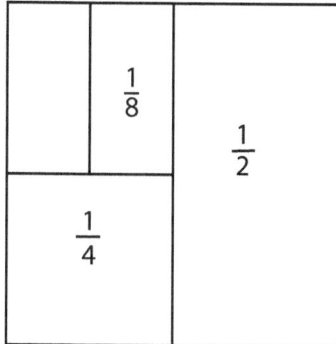

Draw in a few more terms in the box.

Describe what happens to the terms of the sequence, as they go on and on.

Here is another sequence: $\frac{1}{2}, \frac{1}{2}+\frac{1}{4}, \frac{1}{2}+\frac{1}{4}+\frac{1}{8}, \ldots$

Illustrate this sequence in these boxes:

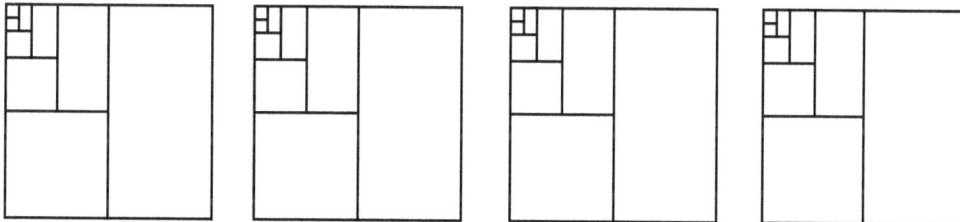

Describe what happens to the terms of this sequence, as they go on and on. If you know how to add fractions, work out the terms and see if they follow a pattern.

1-5.   Days Of The Week

a. What day of the week is today? _____

b. What day of the week will it be 100 days from now? _____

c. 500 days from now, will it be a different day of the week from today?

d. Name a three-digit number for which you know that it will be the same day as today, that many days from now.

1-6. Counting Terms

a. How many terms are in the sequence: 1, 3, 5, 7, ..., 993, 995, 997, 999 ?

b. Find the 1000th term in the sequence: 2, 3, 4, 5, 6, ...

c. Find the 1000th term in the sequence: 2, 4, 6, 8, ...

1-7.   Cumulative Peanuts

Ella the elephant ate 1000 peanuts from Monday through Friday. Each day she ate 10 peanuts more than the previous day. How many peanuts did she eat on Friday?

1-8.   Amoebas

An amoeba reproduces by dividing into two. Suppose it takes a day for an amoeba to eat enough food to be able to divide, and a scientist drops an amoeba into a pond in his back yard on January 1.

a.   Write down the sequence of amoeba population values for January 1, January 2, and so on.

b.   The scientist notices that the pond is full of amoebas on February 10.
     On what date would the pond be full if he dropped two amoebas instead of one on January 1?

1-9. Trina's Sequence Machine

Trina has a sequence machine. She gives it a first number and the machine manufactures the next term, according to these rules:

Rule 1: If the number is less than 10, add 3 to get the next term.
Rule 3: If the number is equal to 10, subtract 5 to get the next term.
Rule 4: If the number is greater than 10, subtract 6 to get the next term.

a. If the first term is 9, what is the 100th term?

b. If the first term is 2, what is the 100th term?

1-10. Julio's Sequence Machine

Julio has a sequence machine. He gives it a first number and it manufactures the next terms according to these rules:

Rule 1: If the number is greater than 8, subtract 7 to get the next term.
Rule 2: If the number is less than or equal to 8, double it to get the next term.

Work out the sequences for various different starting numbers.

First Number	Julio's Sequence
0	
1	
2	
3	
4	
5	
6	
7	
8	
9	

What patterns do you see?

Each sequence is different, because it has a different first term, but are some of Julio's sequences similar to each other, or related to each other?

What's a good way to define *similar sequence* in the world of Julio's sequences?

How To Deal With Sequence Problems

Learn the vocabulary:

Look for patterns such as:

There might be surprises such as:

2. Counting

1-2-3,   2-2-3,   3-2-3, . . .

one potato, two potato
three potato, four

Counting: Notes and Terminology

Counting is probably the first thing you ever learned to do in math. It might seem like the easiest thing to do. But there are many different kinds of counting problems.

Here is one kind of problem: How many rectangles can you count in the following diagram?

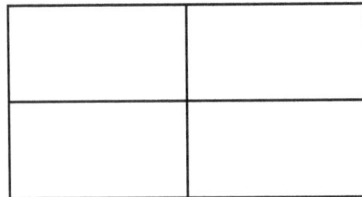

It looks like a diagram made up of four rectangles. But there are more, because for example, the top two rectangles together make up a fifth rectangle. How do you make sure you have counted all possible rectangles?

Think about it. A strategy is described on the next page.

Here is something else to think about. Here are two boxes of sixteen dots.

Box A   Box B

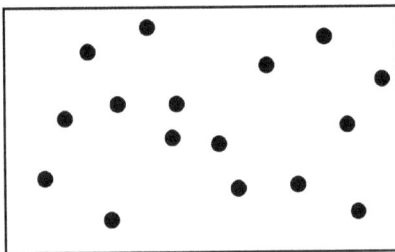   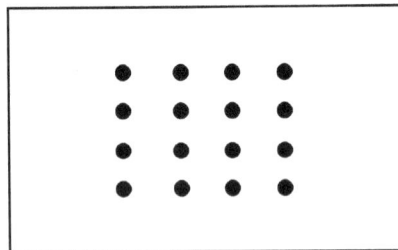

Which set of dots is easier to count, A or B?   _____

Write down some reasons for your answer.

_____

_____

Here is a third kind of counting problem. Suppose you have a coin with a head on one side, a tail on the other. If you toss the coin twice, how many different combinations of heads and tails can you have?

Suppose the first toss gives you a head. The possibilities are: the second toss is also a head, and the second toss is a tail. We can list the two possibilities of head first like this:

HH   and   HT

Now, to count up all the other possibilities, they would each be the same except they would have a T first:

TH   and   TT

What this means is that you can count up the head-first possibilities first, and then double them, knowing that for every head-first possibility, there is a corresponding tail-first possibility. This method is an even better short cut if you try to count all the possible head-tail combinations for three coin tosses. Here are the head-first combinations:

HHH   HHT   HTH   HTT

Without listing further, you can say that the total number of combinations is eight.

But just to make sure, here are all the other combinations:

THH   THT   TTH   TTT

Make sure that there are no other combinations.

Here is a strategy for counting the rectangles in the diagram on the previous page.

Label the smallest rectangles. Every possible rectangle is made up of one or more smallest rectangles. Write down all the combinations.

A	B
C	D

The possible rectangles are A, B, C, D, AB, AC, BD, CD, and ABCD for a total of nine.

2-1.   Counting Triangles

Different triangles can be traced using the lines in the figure given below. How many different triangles can be traced?

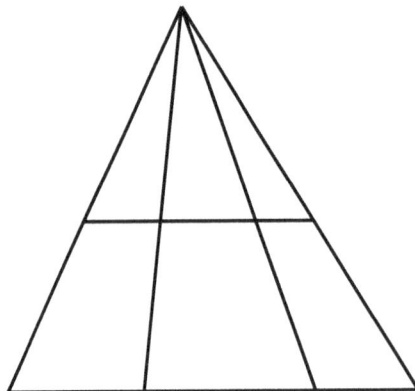

2-2.   Five-Digit Palindromes

How many different five-digit palindromes are there?

Note: a palindrome is a number that reads the same forwards and backwards, such as 232 or 47788774 or 11611).

2-3. Ice Cream Choices

Pat is deciding on her ice cream sundae. She is allowed to have one flavor of ice cream, one syrup, and one topping. Her ice cream flavor choices are: chocolate, strawberry, vanilla. Her syrup choices are: hot fudge, caramel, raspberry, pineapple. Her topping choices are: nuts, sprinkles, gummy bears. How many different possible sundae combinations are there?

2-4. Routes

Here is a map of Alice's house and Bernie's house. Alice lives on the corner of Snowy Street and Windy Street. Bernie lives on the corner of Foggy Street and Rainy Street. How many different routes are there from Alice's house to Bernie's house?

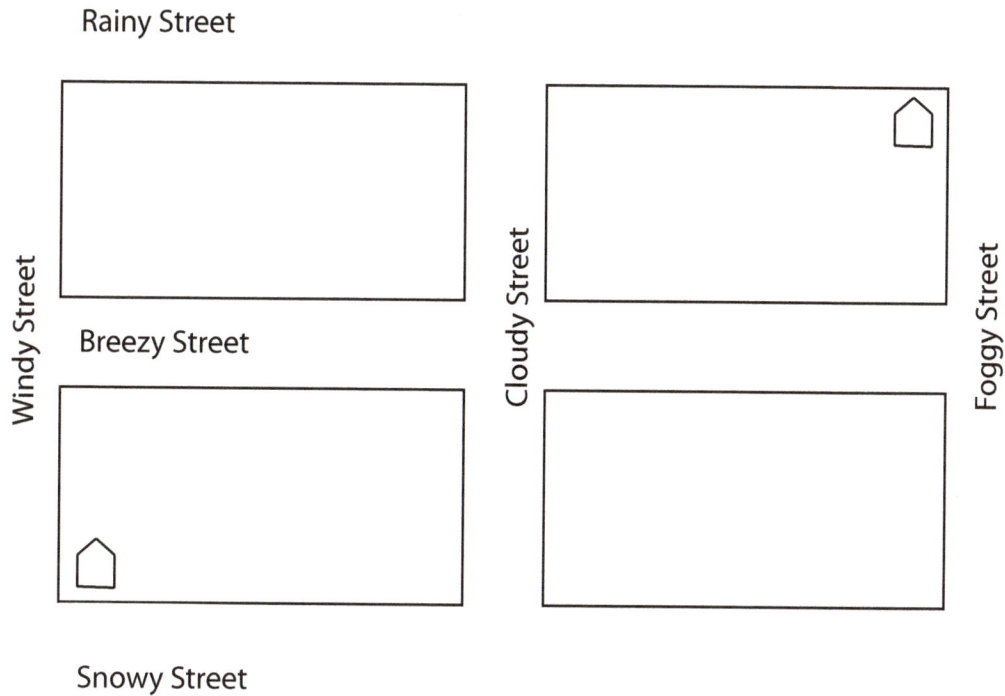

When you answer the first question, you might find that there are different types of routes. How many different *shortest* routes are there from Alice to Bernie's house?

2-5. Page Numbers

Agent Smiley numbers his pages of secret notes consecutively, beginning with 1. This requires a total of 234 digits. How many pages of secret notes does Agent Smiley have?

2-6.   Cover Pattern Counting

The figure on the cover of this book shows 4 patterns of **black** squares.

a. How many sides of squares are on the first (upper left) pattern?

b. How many sides of squares are on the second (upper right) pattern?

c. How many sides of squares are on the third (lower left) pattern?

d. How many sides of little tiny squares are on the fourth (lower right) pattern?

2-7.   Distinct Numbers

How many distinct 4-digit counting numbers can you construct from the digits 3, 0, 0, 7?

2-8. Permutations

Count Rivula has three precious beads: a topaz, an emerald, and a sapphire. The beads can be strung in different order on a necklace that fits around the Count's ruffled collar.

a. How many different bead orders are possible?

b. Countess Rivula gives the Count an additional bead, a ruby. How many different bead orders are possible using all 4 beads?

2-9. Assortments

Henry is allotted 6 cookies from a jar. The jar contains the following types of cookie: chocolate chip, peanut butter, and oatmeal. There are at least 6 cookies of each type in the jar. How many different assortments of cookies could Henry choose?

Note: The following two assortments of cookies are different:
- 2 chocolate chip, 2 oatmeal, 2 peanut butter
- 2 chocolate chip, 1 oatmeal, 3 peanut butter

These two assortments of cookies are the same:
- 3 oatmeal, 2 chocolate chip, 1 peanut butter
- 2 oatmeal, 2 chocolate chip, 1 peanut butter, 1 oatmeal

2-10. Robot Moves

Arlene makes a little robot that moves on a chessboard. The robot can move forward, backward, left, and right, but not diagonally. 5 moves can be programmed at a time. Depending on the sequence of moves, the robot ends up on different destination squares. How many different squares can the robot reach after 5 moves, starting from the upper left?

How To Deal With Counting Problems

Learn the vocabulary:

Look for patterns such as:

There might be surprises such as:

3. Factors And Primes

$2 \times 2 \times 3$

$1001 \div 7$

71, 73

Factors and Primes: Notes and Vocabulary

The **factors** of a whole number are the numbers that divide it evenly. In other words, if you divide a number by one of its factors, the remainder is zero.

The factors of 6 are 1, 2, 3, and 6.

What are the three factors of 9? _____

Some numbers, such as 12, have many factors. Name all the factors of 12:

_____

Name all the factors of 13: _____

13 is an example of a **prime number**. The only factors of prime numbers are themselves and 1.

True or False:

	Statement	T or F?	Example or explanation
1	If 3 is a factor of a number, you can count up to that number by threes.		
2	3 is a prime number.		
3	Find 3 consecutive numbers with a remainder of 2 when you divide by 5. These numbers are all even.		
4	If 2 is a factor of a number, then 4 is also a factor of that number.		
5	If 6 is a factor of a number, then 3 is also a factor of that number.		
6	Some numbers have factors that are prime numbers.		
7	Some numbers have factors that are not prime numbers.		
8	Every number has at least one factor that is a prime number.		

Two obvious prime numbers are 3 and 7.

Name all the prime numbers between 10 and 20: _____

What is the smallest prime number? _____

You might think that the smallest prime number is 1. Its only factors are itself and 1. But the number 1 is not accepted as a prime number. The main reason for this has something to do with the Fundamental Theorem of Arithmetic.

A **theorem** is a mathematical truth that has been proved by at least one mathematician. Many people have proved the Fundamental Theorem of Arithmetic.

Fundamental Theorem of Arithmetic
Every whole number greater than one is either prime or a product of prime numbers. The prime factors of a non-prime number are unique, meaning that if you arrange the prime factors in order of size, there is only one way to do it.

Examples:
7 is prime.
6 is a product of primes: 2 x 3
4 is a product of primes: 2 x 2
11 is prime.
12 is a product of primes: 2 x 2 x 3
12 is a product of many factors, such as 2 x 6 or 3 x 4, but there is *only one way* to express 12 as a product of prime numbers arranged from lowest to largest.

2 x 2 x 3 is called the **prime factorization** of 12.

Prime factorizations are useful for solving problems, because the prime factors of a number are the building blocks of all of the factors of the number. The number 12, for example, has the non-prime factor of 6. But 6 is the product of two prime factors, 2 and 3. So the 2 and the 3 within 12 build up the factor of 6.

Here is a problem that uses the prime factorization idea:

The product of two consecutive whole numbers is 210. What are these two numbers?

Solution: the prime factorization of 210 can be rearranged into any of the factors of 210. So the prime factorization gives us a place to start. 210 is divisible by 2, 3, 5, and 7 (check the divisibility rules).

The prime factorization of 210 is turns out to be 2 x 3 x 5 x 7. We can see that these numbers can be rearranged into 2 x 7 x 3 x 5, which is 14 x 15. Therefore the solution of our problem is: the two consecutive numbers are 14 and 15.

Perhaps you can now see why it's not a good idea to make 1 a prime number. What would happen to the prime factorization of a number if 1 were allowed to be a prime number?

Divisibility Rules

A divisibility rule tells you whether or not a number is divisible by a certain number. Here are some divisibility rules.

Factor	Divisibility Rule
2	Number ends in 0, 2, 4, 6, or 8.
3	The sum of the digits is divisible by 3. Name a 5-digit number divisible by 3:
4	The last 2 digits have to be divisible by 4.
5	Number ends in 0 or 5.
7	1. Double the last digit. 2. Subtract the double from the remaining digits and see if the resulting number is divisible by 7. Example: is 406 divisible by 7? 6 x 2 = 12, and 40 – 12 = 28 which is divisible by 7. So 406 is divisible by 7.
9	The sum of the digits is divisible by 9. Name a 4-digit number divisible by 9:
10	Number ends in 0.
100	Number ends in 00. Name a 6-digit number divisible by 100:

Numbers such as 4 and 10 are **composite** numbers (whole numbers that are not primes). To be divisible by 10, a number has to be divisible by both 2 and 5. Write down some divisibility rules for these composite numbers:

Factor	Divisibility Rule
6	
15	

3-1. Commutativity

What is the value, in simplest terms:

a. $(20 \times 24 \times 28 \times 32) \div (10 \times 12 \times 14 \times 16)$

b. $(26 \times 12 \times 28 \times 15) \div (13 \times 24 \times 14 \times 30)$

3-2.   Factor Counts

The number 6 has four factors: 1, 2, 3, and 6.

a. Which number has more factors, 24 or 28?

b. Which number has more factors, 31 or 33?

3-3. Remainders

What positive whole numbers have a remainder of 4 when you divide them into 32?

3-4.   Large Power Of 2

2^{2012} means the number 2 multiplied by itself 2012 times. It is a very large number. What is its ones digit?

3-5.   Divisibility

How many positive whole numbers less than 100 are divisible by 6 or 21 or by both?

3-6.   Ending Zeros

The number 25! (25 factorial) is equal to 1 x 2 x 3 x ... x 23 x 24 x 25 and is a very large number. It ends with a certain number of 0 (zero) digits.

How many zeroes are at the right-most end? Give reasons for your answer (explain why there are no fewer zeroes, and no more zeroes, than your answer).

3-7.   Consecutive Factors

a. The product of two consecutive numbers is 1806. What are these two consecutive numbers?

b. The product of two consecutive numbers is 1260. What are these two consecutive numbers?

3-8. Factors Of Consecutive Numbers

Name 3 consecutive numbers less than 100, where the smallest is divisible by 7, the next number is divisible by 5, and the largest number is divisible by 3.

3-9.   Remainders With Apples

Jill has some apples. If she puts them in groups of 5, she has 3 left over. If she puts them in groups of 7, she has 2 left over. What is the least number of apples she could have?

3-10. 12345

How many different five-digit numbers can you make using the digits 12345?

How many of these numbers are prime?

How To Deal With Factors and Primes Problems

Learn the vocabulary:

Look for patterns such as:

There might be surprises such as:

4. Related Numbers

once bitten, twice shy

50% off while quantities last, . . .

Four times the GDP

Related Numbers: Notes and Vocabulary

There are many problems about related amounts of things. Some examples:

- Sarah has 2 marbles. George has 5 marbles more than Sarah. How many marbles does George have?

- Marissa drove for 2 hours at 40 miles per hour. How far did she travel?

- Chu can paint 1 garage door in 1 hour. Sven can paint 1 garage door in 3 hours. How long would it take for them to paint 4 garage doors if they work together?

- In the morning Jose spent half his money on a package of batteries, and then at noon he spent two-thirds of what was left on a sandwich. Now he has 2 dollars. How much money did he have to start with?

You might be familiar with the term "word problem." These problems are all word problems. All of these word problems have something in common: they relate one number to another number.

- Sarah's number of marbles is related to George's number of marbles.
- Marissa's driving time and her speed are related to how far she traveled.
- Chu's painting time and Sven's painting time are related to how long it would take them to paint something together.
- Jose's number of dollars, after buying a sandwich and some batteries, is related to how much money he had in the morning.

The problems above can be solved in many different ways.

Technique: Working Backwards
The problem of Jose's money is ideal for solving by working backwards.
At the end of the problem, he has 2 dollars. He had just spent two-thirds of his money on a sandwich and 2 dollars was left, so 2 dollars is the remaining one-third.

Just before he bought the sandwich, he must have had 6 dollars. The 6 dollars is the half left after he bought batteries, so before he bought the batteries, he must have had 12 dollars. And that is how much money he had to start with.

Technique: Read With Care and Use Logic
This basic technique is vital for all related number problems. It's all you need to solve the first problem: *Sarah has 2 marbles. George has 5 marbles more than Sarah. How many marbles does George have?*

In a problem like this, you read for the information that will help you solve the problem. You could ask yourself: What do I know?

How many marbles does Sarah have? 2.
Who has more marbles, George or Sarah? George—so he must have more than 2 marbles.

How many more marbles does George have than Sarah? 5.
The words "more than" mean that to get George's number of marbles, you have to add 5 to Sarah's number. That means the answer is 7.

Technique: Use the Speed, Time, and Distance Relationship

There is a simple relationship between speed, time, and distance traveled. This is what you use to solve the next problem: *Marissa drove for 2 hours at 40 miles per hour. How far did she travel?*

The meaning of "forty miles per hour" is: in 1 hour, 40 miles are traveled. In 2 hours, 40 + 40 miles are traveled. If Marissa drove for 2 hours at 40 miles per hour, she traveled 80 miles.

You might find it useful to remember a formula: the product of the time and the speed is the distance traveled.

Here is another problem that uses the speed, time, and distance relationship: *Fred ran from home to the town of Hayfield, a distance of 12 miles, in 3 hours. What was Fred's average speed?*

The average speed is the total distance divided by the total time. In the case of this problem, Fred's average speed is:

12 miles ÷ 3 hours = 4 miles per hour.

The average speed is the speed that Fred would travel if he traveled at the exact same speed during all 3 hours.

The last simple problem of this type asks for the time: *How long does it take to fly a distance of 1000 miles in an airplane that travels 500 miles per hour?*

The travel time is the total distance divided by the speed. The answer here is 2 hours.

Technique: Use the Rate, Time and Job Relationship
This relationship is the same as the speed, time and distance relationship; it just involves different words. Instead of speed, these problems have a rate; instead of distance traveled, a job is done.

Instead of:
Distance = speed multiplied by time and   time = distance divided by speed

We have:
Number of jobs = rate multiplied by time, and time = number of jobs divided by rate

We use the rate, time and job relationship to solve this problem: *Chu can paint 1 garage door in 1 hour. Sven can paint 1 garage door in 3 hours. How long would it take for them to paint 4 garage doors if they work together?*

This problem presents us with two rates. Chu's rate of door painting is 1 door per hour. Sven's rate of door painting is 1 door per 3 hours, or 1/3 of a door per hour.

The problem asks how long it would take Chu and Sven to paint 4 garage doors if they work together. There are several ways to look at this problem.

One way is to ask: what happens after 1 hour, if Chu and Sven work together? What happens in 2 hours? 3 hours? Is there a pattern? Try this out. Write down what happens:

After 1 hour: _____

After 2 hours: _____

After 3 hours: _____

What do you think the solution is now? _____

Another approach: If Chu's rate is 1 door per hour, and Sven's rate is 1/3 door per hour, what is their combined rate (the rate for the two of them working together)? Think about it. You'd expect their combined rate to be bigger than each of their single rates.

The number of jobs to be done in this problem is 4 (4 doors to paint). Using our rate/time/jobs formula:

Time = number of jobs divided by rate = 4 divided by (combined rate)
See if this method gives you the same answer as the first method.

Technique: Balancing

The idea of a balancing scale can help solve a lot of related number problems. Here's an example: *Mary the baker wants to figure out how much it costs to make a loaf of bread, and how much it costs to make a roll. She knows that 30 cents worth of supplies makes 1 loaf of bread and 5 rolls. And she knows that 60 cents worth of supplies makes 4 loaves of bread.*

We can pretend we have a balance that balances money. If 4 loaves of bread costs 60 cents, we can show this on a balance:

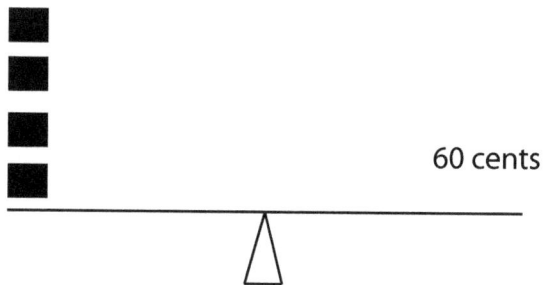

60 cents

If we take away 2 loaves of bread from the left side, what do we take away from the right side in order to balance the scale?

Use this idea to figure out the cost of 1 loaf of bread: _____

Now, 1 loaf of bread and 5 rolls balances with 30 cents. Is there a way to balance *loaves and rolls* with 60 cents? Fill in the sketch:

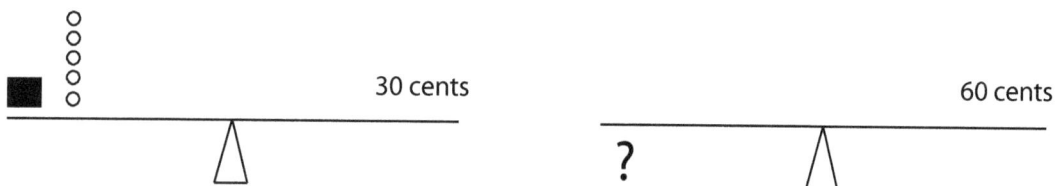

30 cents   60 cents

?

Now you can keep balancing to figure out the cost of 1 roll.

If 1 loaf and 5 rolls balance 30 cents, then 2 loaves and 10 rolls balance 60 cents. We also know that 4 loaves balance 60 cents. Therefore, 4 loaves balances 2 loaves and 10 rolls.

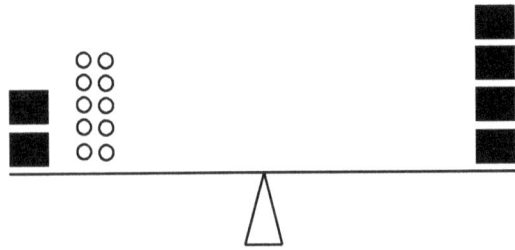

If you take the same things away from each side of the balance, the balance is still kept. Sketch what happens to the above balance if you take 2 loaves away from each side.

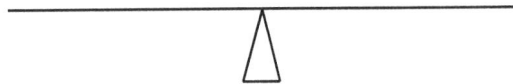

Next, sketch what happens to the balance if you take half of each side away.

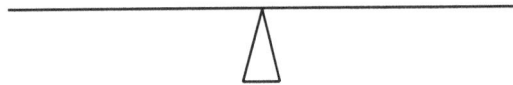

Re-sketch the previous balance, replacing the loaf with the amount of money it is worth.

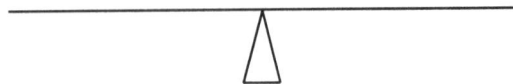

How much money does 1 roll cost? _____

Technique: Variables and Equations

A variable is a symbol for a quantity. A variable gives you a short way to write down a quantity. For example:

Herman the centipede has twice as many legs as Sandra the caterpillar.

If we let the variable X stand for the number of legs Sandra has, then the number of legs that Herman has is X + X.

In math, we say: Let X = the number of legs that Sandra has.
We can also say: Let Y = the number of legs that Herman has.

Then we can write an equation: X + X = Y

X + X means "two X's" and in math, we write that as 2X.

How do you think we write X + X + X? _____

The equation relating Herman's and Sandra's legs can be written: 2X = Y

If Sandra has 10 legs, how many does Herman have? _____

If Herman has 14 legs, how many does Sandra have? _____

If you have a problem like the Loaf/Roll problem, you can use variables and equations. In fact, the equation technique and the balancing technique are the same. Instead of a balance, you write an equation. Every diagram of the balance can be converted to an equation. For example:

Balance	Equation
60 cents	Let L = cost of 1 loaf L + L + L + L = 60 or: 4L = 60

Equations work like balances. Just like on a balance, you can take away half from both sides of an equation and you still get an equation.

If: L + L + L + L = 60 = 30 + 30

Then, taking half away from each side: L + L = 30

Another example from the previous problem:

Balance	Equation
30 cents	Let L = cost of 1 loaf Let R = cost of 1 roll L + R + R + R + R + R = 30 or: L + 5 R = 30

You can double equations:

If: L + 5R = 30

Then, doubling: L + 5R + L + 5R = 30 + 30

When you add, you can rearrange the items being added:

L + L + 5R + 5R = 30 + 30

And you can add up matching items:

2L + 10R = 60

Try converting this balance to an equation:

Balance	Equation
	Let L = cost of 1 loaf Let R = cost of 1 roll ?

Try solving the Loaf/Roll problem entirely with equations, following the solution of the balances.

4-1.   Balancing Fruit
One apple weighs as much as a banana and a tangerine together.

a. Draw a balance scale in a balanced position with an apple on one side.

Answer the following questions, if possible. Some of these questions cannot be answered—there isn't enough information about all the fruits. If a question cannot be answered, say so and think about why.

b. Does the apple weigh more than the banana?

c. Does the banana weigh more than the tangerine?

d. Does the apple weigh more than two bananas?

e. Does the tangerine weigh less than the apple?

f. Does the banana weigh less than the apple?

g. Does the apple weigh more than two tangerines?

h. Do two tangerines weigh less than two apples?

4-2. Marbles

Shirley gave half of her marbles to Harry. Harry gave 3 marbles to Dion, and one-third of what he had left to Sharon. If Sharon has 4 marbles, how many did Shirley have?

4-3.   Car Factories

There are 3 factories in the province of Atheneum. The factory in Smallville builds 2 cars per month. The factory in Bigville builds 4 cars per month, and the factory in Rightville also builds 4 cars per month.

a. How long does it take the Smallville factory to build 10 cars?

b. How long does it take the Bigville factory to build 10 cars?

c. How many cars per month are built in the entire province of Atheneum?

d. How long would it take for Atheneum to produce 200 cars?

4-4. Chocolates And Lollipops

4 chocolates and 1 lollipop cost 48 cents.

1 chocolate and 4 lollipops cost 72 cents.

What is the cost of 1 chocolate and 1 lollipop?

4-5.   Forks, Spoons, And Steel

From 4 ounces of steel, an artisan can make 1 fork and 1 spoon.
From 10 ounces of steel, the artisan can make 3 forks and 1 spoon.

a.   How many ounces of steel go into 1 fork?

b.   How many ounces of steel go into 1 spoon?

4-6.   Race To The Middle

John and Paul go toward one another in a straight line. John walks at 2 miles per hour. Paul jogs at 4 miles per hour. They meet in 30 minutes. How far apart were they when they started?

4-7. Dog Supplies

Sally has 4 dogs. She has enough dog food to last 30 days. She agrees to take care of 2 more dogs while her friends go on vacation. Assuming each dog eats the same amount of food each day, how many days will Sally's dog food last?

4-8.   Number Trick

Here is a number trick: Choose any positive whole number. Add 4 to it. Multiply the result by 2. Subtract your original number. Then subtract 5. Subtract the original number once more. Do this trick a few times, with different starting numbers. You end up with a certain number every time.

a. What is the number you end up with?

b. Explain why the trick works.

c. Does the mystery number have to be whole for the trick to work? Does it have to be positive?

4-9.   Picking Peaches

An adult takes 3 hours to pick all the peaches from a tree. A child takes 7 hours to pick the same tree. How long would it take the adult and child to pick all the peaches from this tree, if they work together?

Give your answer in hours and minutes.

4-10. Filling A Draining Tub

A tub takes 4 minutes to fill, 5 minutes to drain. How long does it take for the tub to fill if the tap is running and the drain is open?

Professor Bear says: Don't do this at home.

How To Deal With Related Number Problems

Learn the vocabulary:

Look for patterns such as:

There might be surprises such as:

5.   Statistics

Four out of five mechanics recommend . . .

January rainfall, mm

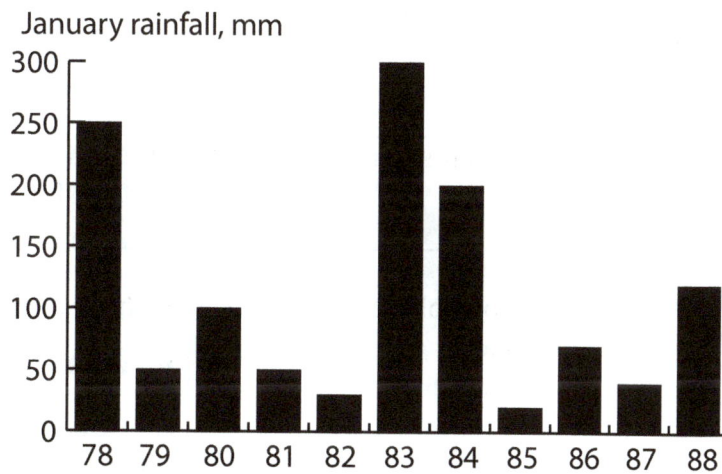

Statistics: Notes and Vocabulary

Data is information you collect by taking measurements or counts.
The word "data" is plural and refers to many measurements or counts.

Some examples of data sets:
- The heights of all students in a particular classroom.
- The number of migrating birds that rest on a lake each day in the autumn.
- The temperature outside of your house measured every hour from 8:00 AM to 8:00 PM.

A **statistic** is a number or value calculated or derived from a data set. Here are some examples:

Data Set	Description of Data	Statistic
8, 9, 9, 9, 10	Ages of kids in a math club	**sample size**: 5 (there are 5 data points in the data set)
8, 9, 9, 9, 10	Ages of kids in a math club	**mean**: $\dfrac{8+9+9+9+10}{5}$ = 45/5 = 9 = average age
52, 53, 58, 60, 61	Heights of kids in math club, in inches	**median**: 58 (data point in the middle)
0, 0, 1, 1, 1, 2, 3, 3, 3, 4	Number of siblings of kids in school play	**sample size**: 10 **median**: $\dfrac{1+2}{2}$ = 1 and ½ (if the sample size is even, the median is the average of the middle two values)
khakis, jeans, chinos, jeans, jeans	Type of trousers worn by kids in math club	**mode**: jeans (most frequently occurring data point)
goldfish, cat, cat, dog, dog	Pets owned by kids in math club	**mode**: cat **mode**: dog (data set has two modes)

Mean

The average of n numbers is the sum of the numbers divided by n.
The average is also called the **mean**.

What is the mean of 1,2, and 3? _____

What is the mean of 0, 1, and 2? _____

Do you think it's possible for the average of 5, 5, 6, 7, 7, 8, 9 to be more than 10?
What about less than 6? (Try to answer without calculating, and explain.)

_____

Median

If a data set consists of numbers, the median is the middle value.

In a list with an odd number of numbers, the median is the middle value. In the data
set 3, 5, 7, 10, 11, the median is 7.
What is the median of: 1, 1, 1, 2, 55, 56, 99? _____
Notice that the median is *not* the half-way point between the maximum and
minimum values. The half-way point between 1 and 99 is 50.
What's the median value of 100, 101, 199, 200, 201, 399, 900? _____

In a list with an even number of data points, the median is the average of the two
middle values.

What is the median of 2, 3, 4, 6, 9, 18? _____

What's the mean of 2, 3, 4, 6, 9, 18? _____

If the numbers in the data set have distinct (unequal) values, the median divides the
data set into two halves: a half below the median, and a half above the median. If the
data values are not distinct, this does not happen. Compare the medians of:
1, 1, 1, 1, 1, 1, 1, 7, 8, 9: the median is: 1. Less than half the data is above 1.
1, 3, 4, 6, 7, 9, 11, 12, 17, 20: the median is 8. Half the data is below 8, half above 8.

Mode

The mode of a list of numbers is the most frequently occurring number in the list.
The mode of 0, 0 , 0, 1, 2, 3, 3 is 0.
What about the mode of: 3, 3, 4, 4, 5, 5? This has three modes: 3, 4, and 5.

Write down a mode question for a friend:

What is the mode of: _____

5-1.   What's A Statistic?

In the left column are statistics. In the right column are data sets. Match each statistic with a suitable data set.

Statistics	Data Sets
One out of three data points is an apple.	The age, in years, of each human being alive on Earth right now.
The mode is 10.	The favorite hat worn by jazz dancers in the late 1980s.
The mean is greater than 35.	The amount of time each book was checked out from the Vancouver Public Library, in the year 2000.
The data values are between 0 and 130.	The number of salmon that swim to the mouth of the Skeena River each summer.
The median is 19 days.	The ages of fifth graders in New York during the month of October in each year from 1960 to 2012.
The mean is about 5 million.	The favorite fruit of each astronaut who rode a NASA space shuttle.
There are two modes: the fedora and the porkpie.	The average retirement age of police officers in Elvewellyn.

5-2.   Page Number Statistics
Paul read a book with 37 pages. The pages are numbered 1, 2, 3, ... , 35, 36, 37.

a.   What is the median page in Paul's book?

b.   How many digits are used to number the 37 pages in Paul's book? Hint: the number of digits used to number pages 1 through 12 is 15.

c.   Paul reads a second book that has page numbers that use up 77 digits. What is the page number of the median page in the second book?

5-3. Related Means

The mean (average) of four numbers is 17. The largest number of the four is 23. What is the mean of the remaining three numbers?

5-4. Mouse Data

Ten children try to measure the weight of a wiggly baby mouse. They don't get the same measurements. One child measured the weight as 2 grams. Three children measured the weight as 4 grams. Here is the measured data:

Weight measurement of wiggly baby mouse	Number of kids who got that measurement
2 grams	4
4 grams	3
5 grams	2
6 grams	1

a. Sketch the frequency distribution bar graph for the weight data.

b. Calculate the mean, the median, and the mode of the weight data.

c. What do you think is the best estimate of the true weight of the wiggly baby mouse? The mean? The median? The mode? Explain your conclusions.

5-5.   Five Mystery Numbers

A set of five positive whole numbers has a mean of 4, a median of 4, and a single mode of 6. What are the five numbers?

5-6.   Family Reunion

Thirty people attend a family reunion. The average age is 17 years. There are 12 girls, 10 boys, and 8 adults. If the average age of the girls is 10 and the average age of the boys is 7, what is the average age of the adults?

5-7.   Averages Of Subsets

In a set of seven numbers, the average of the first four numbers is 6, and the average of the last four numbers is 10. If the average of all seven numbers is 8, what is the number common to the first four numbers and the last four numbers?

5-8.   Rainfall

Here are annual rainfall bar graphs for two different locations, A and B.

Annual Rainfall Bar Graph A

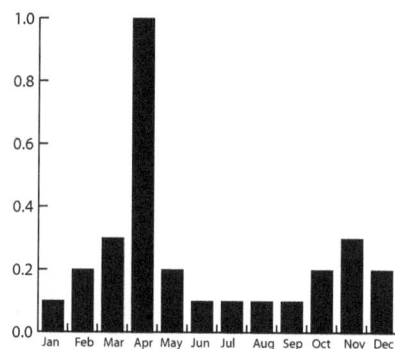

Annual Rainfall Bar Graph B

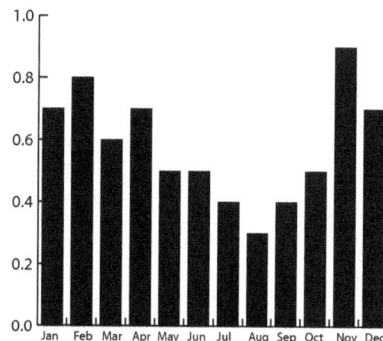

a.   Which location has the higher average rainfall, A or B?

b.   Which location has the highest rainfall in a month (in any month of the year)?

c.   Which location has the lowest rainfall in a month?

d.   Which location, A or B, is most likely to be a desert?

e.   Which location has the most **even** distribution of rainfall? Use numbers to justify
     your answer.

5-9.   Truth And Statistics

Compare and contrast statement A and statement B:

A   Out of 100 dentists surveyed, 80 dentists said they preferred Whizzy-Floss to all other brands of dental floss.

B   4 out of 5 dentists recommend Whizzy-Floss.

5-10. Statistics Problem 10

Here is a small data set:

7
7
8
8
10

a. Calculate the mean, median, and mode.

b. How much would you have to change the first data point to lower the mean by 1 unit? What would happen to the median and mode if you lower the first data point?

c. How much would you have to change the last data point to raise the mean by 1 unit? What would happen to the median and mode if you raise the last data point?

d. Do you see a pattern to the answers of (b) and (c)? What if, instead of five data points in the data set, there are 100 data points—how much would the data values have to change to shift the mean up by one unit?

How To Deal With Statistics Problems

Learn the vocabulary:

Look for patterns such as:

There might be surprises such as:

6. Logic

If wishes were horses, then ...

Logic: Notes and Vocabulary

Logic is the study of what is true and what makes sense. A frequent question in logic is: how do you assemble small truths into big truths? In other words, can a few clues (assorted little truths) lead to the solution of a mystery?

In math, a basic building block is number. We often work with numbers.
In logic, a basic building block is a **statement**. A statement is a sentence that can be true, or false, but not both.

Statements:
      Even numbers are divisible by two.
      Even multiplies of five end in zero.
      My brother lives on Mars.

Not statements:
      Hey, Bob!
      I am lying.

Clue Puzzles

There are many puzzles where you have a few true statements about something, and the goal of the puzzle is to figure out everything from a few true statements.

Example: Melissa, Susan, and Phil live in the same town. One of them is 10, one is 11, and one is 12. Phil is the same age as Susan's younger brother. Melissa is one year older than Phil. What are the ages of Melissa, Susan, and Phil?

A good way to solve this puzzle is by making a chart of all the possible combinations of names and ages. Then cross off combinations that are ruled out by the clues:

	Melissa	Susan	Phil
10			
11			
12			

One clue says: *Phil is the same age as Susan's younger brother*. This means that Phil is younger than Susan, and therefore Phil cannot be the oldest person. Also, Susan cannot be the youngest person. From this clue alone we can cross off two possibilities:

	Melissa	Susan	Phil
10		X	
11			
12			X

The next clue is: *Melissa is one year older than Phil.* This means that Melissa cannot be the youngest person. Here is the updated chart:

	Melissa	Susan	Phil
10	X	X	
11			
12			X

Because one of the three people must be 10, the 10 year old must be Phil. Here is the update of the chart:

	Melissa	Susan	Phil
10	X	X	√
11			X
12			X

Since Melissa is one year older than Phil she must be 11, and then Susan must be 12.

Another example of a clue puzzle is a Number Place puzzle, also known as Sudoku. Here is a simple Number Place puzzle:

2		4	
3		2	
	3		

The rules of this puzzle are that each row, each column, and each outlined two-by-two square must contain the digits 1, 2, 3, and 4.

Because, for example, the upper left square has to contain all four numbers, the two empty spots must contain 1 and 4. Because 4 is in the top row already, the lower spot must contain the 4:

2	1	4	
3	4	2	
	3		

You can now proceed to fill in the rest of the spots. In the first row, the missing digit is 3; in the second row, it is 1, and so on.

You can see how these puzzles are chains of logically related true statements:
Statement: There are a 2, 1, and 4 in the upper row.
Statement: Each row must have all four digits, 1, 2, 3, and 4.
Statement: Therefore, the upper row must have a 3 in the right-most spot.

Put another way, solving the puzzle means making a chain of logically related true statements. The chart and the Number Place grid are ways to organize the clues so that the solution becomes easier to see.

Cipher Puzzles
A cipher is a translation of letters into numbers or numbers into letters. Using logic from arithmetic, you can solve simple cipher puzzles.

Example: The letters U, R, and X represent three different single positive digits. Find out what the digits are if they obey this sum:

$$\begin{array}{r} U \\ +R \\ \hline RX \end{array}$$

Solution: Because U and R are single digits, the sum RX has to be less than 20. That means that R has to be 1. Now we can translate part of the sum:

$$\begin{array}{r} U \\ +1 \\ \hline 1X \end{array}$$

This means that U + 1 = 1X
The only one-digit number that becomes a two-digit number if you add 1 to it is 9 (any other digit is too small). So U must be 9. And then X must be zero (0).

Venn Diagrams

Here is an example of a logic problem:

At Smithson School, there are 50 fifth graders. They all wear some form of school clothing: a hat or a vest. 25 fifth graders wear school hats. 35 fifth graders wear school vests. How many fifth graders wear both hat and vest?

Solution 1: the total number of students is 50. 25 + 35 = 60. So there must be some students who wear both hats and vests. 60 – 50 = 10, so there must be 10 students who wear both hats and vests.

Solution 2: Venn diagram.

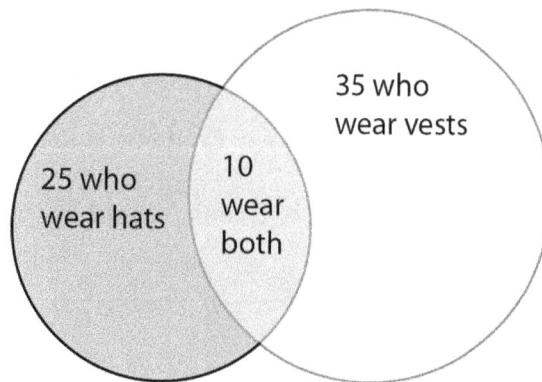

The Venn diagram above shows a circle for the 25 who wear hats, and an overlapping circle for the 35 who wear vests. The overlap of the two circles represents students who wear both hats and vests.

You can see that the group of 25 students is made up of those who wear hats but not vests plus those who wear both hats and vests. You can write an equation:

25 = hats-only fifth graders + fifth graders who wear both

Write a similar equation for the group of 35 students who wear vests:

_____

Write a similar equation for the total group of 50 students:

_____

Debunking

It is often harder to prove that something is true than to prove that something is false. All it takes to prove a statement is false is one example of falseness—a **counterexample**.

It is worthwhile to practice the ability of coming up with good counterexamples. Here are a couple of outlandish statements. Come up with counterexamples:

Statement: If a number is divisible by 2, it is also divisible by 62.

Counterexample: _____

Statement: If a number N is divisible by 7, then N+1 is divisible by 8. (Notice that this statement is true if N=7.)

Counter example: _____

Here are some more statements that are false. Come up with counterexamples:

Statement: There are no prime numbers between 30 and 39.

Counterexample: _____

Statement: If you measure a rectangle's area in square inches and its perimeter in inches, it is not possible for the perimeter to be a larger value than its area.

Counterexample: _____

Statement: If an apple weighs the same as a banana and a tangerine, then the apple weighs the same as two bananas.

Counterexample: _____

6-1.   Adding Mystery Digits

Three digits, A, B, and C obey this sum:
$$\begin{array}{r} AAA \\ +BBB \\ \hline AAAC \end{array}$$

What are A, B, and C?

6-2. Multiplying Mystery Digits

Two digits, A and B, obey this product:

$$
\begin{array}{r}
ABBA \\
\times\ \ AA \\
\hline
AABAA
\end{array}
$$

What are the digits A and B?

6-3. Number Place Puzzle

Fill in the Number Place puzzle. The rules are: each row contains the digits 1, 2, 3, 4; each column contains 1, 2, 3, 4; and each outlined 2x2 square contains 1, 2, 3, 4.

			1
	4		2
	3		

6-4.   Cross Sum Puzzle

Here is a Cross Sum puzzle. You fill it in with numbers that add up to the white numbers on the left, if the numbers are along a row, or add up to the white number above, if the numbers are along a column. No digits can be repeated.

Examples:

The empty squares above can be filled with 1 and 3, 3 and 1, but not 2 and 2.

The empty squares at left can be filled with 7 and 9, 9 and 7, but not 8 and 8.

Some useful sums:

$1 + 2 = 3$ $1 + 3 = 4$

$1 + 2 + 3 = 6$ $1 + 2 + 4 = 7$

$1 + 2 + 3 + 4 = 10$ $1 + 2 + 3 + 4 + 5 = 15$

$7 + 9 = 16$ $8 + 9 = 17$

What is special about all of these particular sums? (Compare with the possible ways to sum up to 5, or 8, or 9).

6-5.   Who Made What Jam?

Penny, Lucy, Max, and Carl each made jam. The jams they made, in no particular order, are strawberry, raspberry, peach, and fig.

Penny and Carl made berry jams.
Max is allergic to peaches.
Neither Lucy nor Penny like raspberries.

What type of jam did each person make?

6-6.   Who Laid Green Eggs?

Mera, Ciela, and Terra are three chickens. They each lay eggs of a particular color. One day they laid a total of 10 eggs. Some eggs are white, some are brown, and some are green.

Ciela laid the most eggs by 1.
There are equal numbers of white eggs and green eggs.
Mera's eggs are green.

How many eggs of each color did each chicken produce?

6-7.   Implications And Converses

An **implication** is a special kind of statement. It is an "If ... then ..." statement.

Examples of implications:
If Willie sings, then Willie is happy.
If Horace makes a cake, then the kitchen needs cleaning.
If it rains, then Appleton is happy.

The converse of an implication is a related "If ... then ..." statement where the first and second parts are switched. Here are the converses of the example implications above:
If Willie is happy, then Willie sings.
If the kitchen needs cleaning, then Horace makes a cake.
If Appleton is happy, then it rains.

You can see from these examples that implications might be quite different from their converses. And if the implication is true, the converse might be false. Here are some mathematical implications. State whether they are true or false, and write down the converse statement. State whether the converse is true or false.

Implication	True or False?	Converse	Converse True or False?
If a number is even, then the ones digit of the number is either 0, 2, 4, 6, or 8.			
If a number is prime, then the number is not divisible by 4.			
If the remainder of a number is 1 when you divide it by 5, then the number has a ones digit of 1.			

6-8. Debunking Practice

Here are some statements. Determine if they are true or false. If a statement is false, give a counterexample.

Statement	True or False?	Counterexample if False
Any multiple of 10 is divisible by 5.		
Any multiple of 5 is divisible by 10.		
All primes except 2 are odd.		
The largest number is googol.		
If good numbers are divisible by 3, and bad numbers are divisible by 17, then a number cannot be both good and bad.		
If X is a number that is 5 less than another number Y, then X is more than half of Y.		

6-9.   Class Pets

A teacher recorded some data about the pets owned by children in his classroom.

15 children have dogs.

18 children have cats.

8 children have fish but no dogs and no hamsters.

6 children have hamsters but no cats and no fish.

3 children have hamsters and dogs.

7 children have both dogs and cats.

4 children have fish and cats.

How many children are there in the class?

6-10.   Twin Primes

Twin primes are consecutive prime numbers that differ by 2. The numbers 3 and 5 are twin primes. Another set of twins is 11 and 13.

a.   Find all of the twin primes less than 50.

b.   3, 5, and 7 are "triplet" primes, three consecutive primes that increase by 2. Are there any other sets of triplet primes? Explain why or why not.

c.   Can you explain why the number between a pair of twin primes is always divisible by 6? (With the one exception of 4.)

How To Deal With Logic Problems

Learn the vocabulary:

Look for patterns such as:

There might be surprises such as:

7.   Area And Perimeter

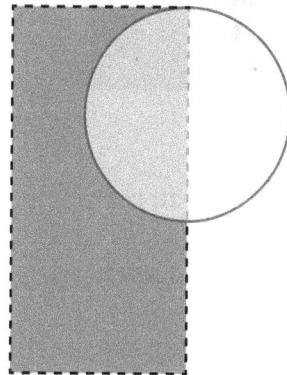

Area and Perimeter: Notes and Vocabulary

Length is the one-dimensional size of something. You can measure length with a ruler.

Area is the two-dimensional size of something: how far does it spread out in two directions. People want to know areas of carpets, lawns, cloth, countries. Here are some questions to think about:

How would you measure the area of a tiled floor?
How would you measure the area of a house window?
How would you measure the area of a car windshield?

What might an "area ruler" look like?

One approach to measuring areas is to cover them with smaller units of a known area. The easiest area to measure is the area of a rectangle, because you measure the length and width of the rectangle with a ruler, and then multiply the length by the width to get the area.

Each of the rectangles below is labeled with its dimensions in Larks.

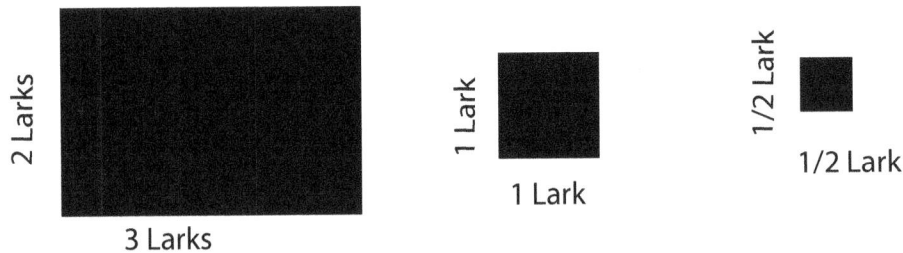

What is the area of each rectangle, in square Larks?

_____ square Larks, _____ square Lark, _____ square Lark

Is the area "bigger" than the length and width? Why or why not?

You can see from the three rectangles above, sometimes the size of the area is larger than either length or width, sometimes it is the same, and sometimes it is smaller.

The question of whether the area is "bigger" than the length is an example of a badly posed question. Areas and lengths are different measurements. Notice that the area is measured in **square Larks**, and the length in **Larks**. You can't compare measurements taken in two different types of units.

Another thing to notice, however, is that because rectangular areas are products they have a strong connection to multiplication and factors. The length and width of a rectangle are factors of its area.

Many rectangle problems involve multiplication and factors, and many multiplication problems can be solved using rectangles.

A final observation: the quirks of multiplication apply to areas. Sometimes when you multiply two numbers, you get a number that is bigger than the original numbers. Sometimes the answer is much smaller. What is the area of a rectangle that measures 0.01 Lark by 0.001 Lark?

Perimeter is the distance around an area. It is a one-dimensional size that you can measure with a ruler.

What is the perimeter of each rectangle on the previous page?

_____ Larks, _____ Larks, _____ Lark

7-1.   Square Tiles

Jonie has 24 square tiles, one-inch by one-inch. With the tiles she can make rectangles.

a.   Whatever she makes, if she uses all of her tiles with no overlapping, the area is

_____.

b.   How many rectangles of different dimensions can she make? The dimensions are the length and width of the rectangle. For example, she can make one rectangle that is 3 tiles wide and 8 tiles long. What are the others?

c.   What is the largest perimeter rectangle she can make?

d.   What is the smallest perimeter rectangle she can make?

7-2. Folded Paper Part One

A square of paper of area 64 square cm is folded in half to make a rectangle. What is the perimeter of the rectangle?

7-3.   Folded Paper Part Two

A square piece of paper is folded in half to make a rectangle. The perimeter of the rectangle is 30 cm. What is the area of the square?

7-4.   Rectangular Tiles

Edgar has three rectangular tiles that are 1 inch wide and 2 inches long.

a. How many different ways can he arrange them in a rectangle of area 6 square inches? Assume that the tiles can be either vertical or horizontal when arranged.

b. What perimeters are possible for Edgar's 6-square-inch rectangles?

c. Edgar's rectangles fall into different groups according to symmetry. Two rectangles are in the same symmetry group if it is possible to move one on top of the other and they match shape exactly. What are the different symmetry groups?

7-5.   Changing Areas

Ralph has 9 tiles of area one square inch. He arranges them in a 3x3 square of perimeter 12 inches.

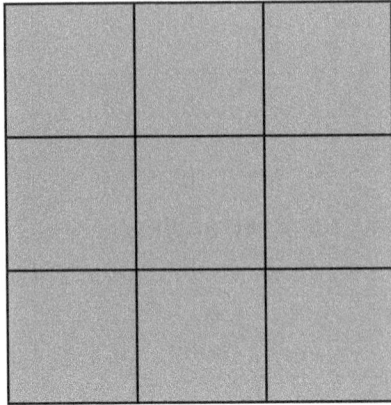

Ralph removes two tiles and gets the area below.

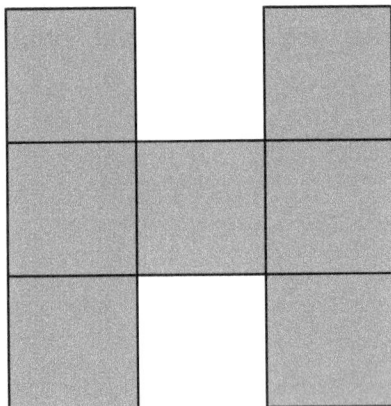

a. What is the area of the second figure?

b. What is the perimeter of the second figure?

7-6.   Circumference

Circle A

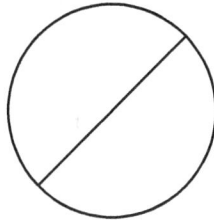

Circle B

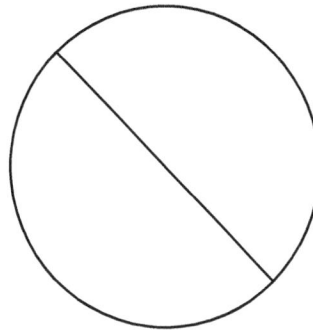

Circle C

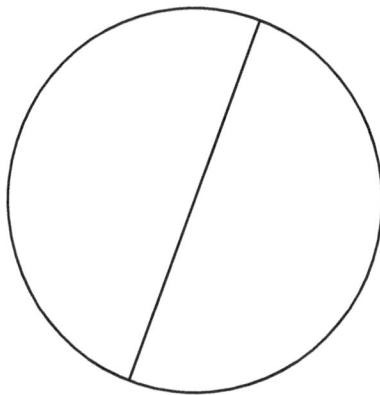

Circle D

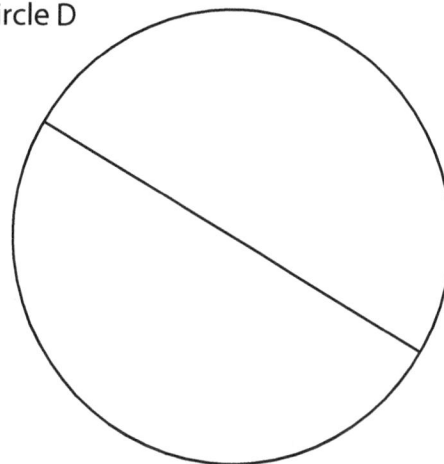

Measure the perimeters and diameters of Circle A, Circle B, Circle C, and Circle D. The perimeter of a circle is also known as the circumference. The diameter of a circle is the widest distance across the circle. Record your data in the table below. Hint: use string to measure the circumference.

Circle	Circumference	Diameter	Circumference ÷ Diameter
A			
B			
C			
D			

What do you notice about the circumference ÷ diameter?

7-7.   Circles Vs. Squares

Imagine a square. You can draw two circles related to the square:

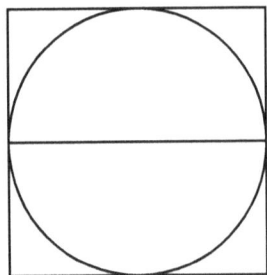

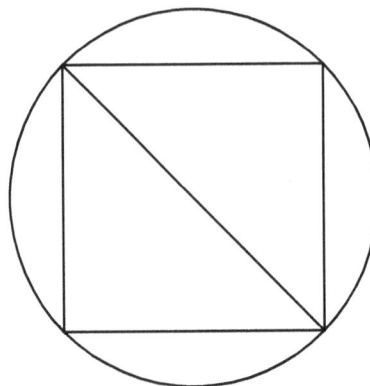

Diameter = side length   Diameter = diagonal length
Circle is smaller   Circle is bigger

In the first case, the circle is smaller (in area) than the square. In the second case, the circle is larger in area than the square. There is a famous ancient problem called "Squaring the Circle" where mathematicians looked for an easy way to find a square and a circle with the same area. They looked for a relationship between the length of the side of the square and the diameter of a same-size circle. This problem turned out to be more difficult than anyone expected.

In the previous problem, you measured the circumference and diameters of four circles. It turns out that for *every* circle, the circumference ÷ diameter is equal to the same number, a special number a little larger than 3 called π.

Using π, we have easy formulas for the circumference (perimeter) and area of circles.
The circumference is π x Diameter, or πD.
The area is π x Radius x Radius, where the Radius is half of the Diameter.
The area formula is also written πR^2.

a.   If a circle has circumference equal to 12π cm, what is the area of the circle in terms of π?

b.   If a square has perimeter equal to 12π, what is the area of the square?

c.   Which has the larger area, the circle with perimeter 12π, or the square with perimeter 12π?

7-8.   Bicycle Wheel

If an 80 cm diameter bicycle wheel turns 2 full revolutions per second, how fast is the bicycle moving in kilometers per hour?

Information you need:
a. The circumference of the wheel is π x diameter.
b. If the wheel turns 2 full revolutions per second, how many circumferences does it travel per second?
c. The speed is the distance traveled ÷ time.
d. There are 100 centimeters in 1 meter.
e. There are 1000 meters in 1 kilometer.
f. There are 3600 seconds in 1 hour.

7-9. Triangle Areas

Inside every rectangle are two triangles. They are called right triangles because one of the angles in the triangle is a right angle. The area of a right triangle is easy to calculate. It's half of the area of the rectangle that the right triangle sits in.

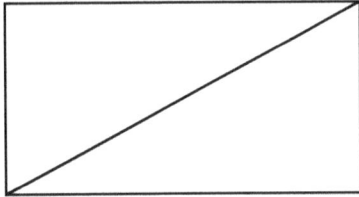

This idea extends to the area of any triangle. If a random triangle is not a right triangle, then you can always rotate it so that one of the sides is horizontal, and the highest point is over the base.

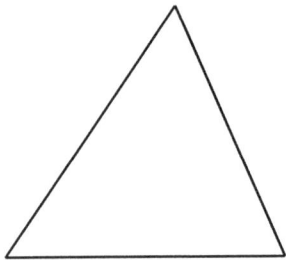

Show that the area of the triangle below is ½ x Base x Height. Use the two rectangles shown.

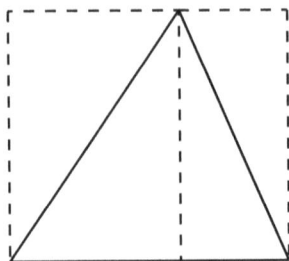

7-10. Cover Squares
On the cover of this book are four diagrams of squares within a square. In the first diagram, there are 5 squares, in the second, 25, and so on.

a. What is the total area of the squares in each diagram? (Four answers)

b. What is the total perimeter of the squares in each diagram? (Four answers)

How To Deal With Area and Perimeter Problems

Learn the vocabulary:

Look for patterns such as:

There might be surprises such as:

8. Probability

10% chance of rain

certainty

coins, dice, weather

Probability: Notes and Vocabulary

An **event** is something that could happen. The **probability** of an event is a number in the range from 0 to 1 that expresses the likelihood of the event happening.

If the probability is 1, the event will definitely happen.
If the probability is 0, the event would never happen.
If the probability is somewhere between 0 and 1, the event might happen. The closer the probability is to 1, the more likely the event will happen.

The probability is this fraction: $\dfrac{\textit{number of ways event could happen}}{\textit{total number of events}}$

If you flip a coin once, there are 2 possible events: it comes up Heads or it comes up Tails.

Coming up Heads is 1 event. The probability that it comes up Heads is $\dfrac{1}{2}$.

The set of all possible events is called the **sample space**. In the case of flipping a coin once, the sample space is the set {comes up Heads, comes up Tails} which we can abbreviate as {H, T}.

One important part of probability is therefore counting. The number in the denominator of the probability is the number of objects in the sample space, so you need to count them.

Example: If you flip two coins, what is the sample space?

Answer: {HH, HT, TH, TT}. The sample space is the set of all possible results of flipping two coins. The size of this sample space is 4.

Another example: if you toss one 6-sided die, what is the sample space?

Answer: {1, 2, 3, 4, 5, 6} The size of this sample space is 6.

If you flip two coins, here some possible events and their probabilities:

Knowing the sample space makes it easier to calculate probabilities.

In the case of flipping two coins, the following table shows some events and how to calculate their probabilities.

Event	Probability of Event, given sample space of {HH, TH, HT, TT}
Two Heads	The number of ways HH occurs in the sample space is 1. Therefore: P(HH) = ¼
One Head, one Tail	There are two ways to achieve one Head and one Tail: HT, TH. P(HT or TH) = 2/4 = ½
Two Tails	There is one way to achieve two Tails: TT P(TT) = ¼
At least one Tail	There are three ways to achieve at least one Tail: HT, TH, TT P(HT or TH or TT) = ¾
Both coins disappear and go to Jupiter	This is so unlikely to happen that it is not in our list of possible events in the sample space. P(disappearing and going to Jupiter) = 0/4 = 0

8-1.   Sample Space

Each of these probability questions has a related sample space. What is the sample space? What is the size of the sample space?

a. Mrs. Cumberbatch is going to buy a new car. She'll choose the color at random. The available colors are: red, silver, white, and black. What is the probability she'll get a silver or black car?

b. What is the probability that the person standing closest to you was born on Thursday?

c. What is the probability of tossing three Heads in a row (with a fair coin)?

d. A computer generates numbers at random from the four digits 4, 5, 6, and 7. What is the probability that it generates the number 7654? (Each digit is used once.)

e. Mrs. Cumberbatch went back to the car store. She found out she can get one of two kinds of tires: radial, or all-weather, and one of three kinds of window tinting: gray, aqua, or taupe. What is the probability she gets all-weather tires and taupe tinting?

8-2.   Gumballs

Mr. Grumby wants to buy four gumballs of the same color from a gumball machine. The machine contains 6 yellow, 7 red, and 8 blue gumballs. How many gumballs does Mr. Grumby have to buy to be certain he has four gumballs of the same color?

8-3.   Sack Of Balls

A sack contains balls that are identical except for color. There are 6 red balls, and the remaining balls are all green. If the probability of drawing a red ball at random from this sack is ¼, how many green balls are in the sack?

8-4. Six-Sided Die
A six-sided die is thrown twice.

a. What is the probability that the sum of the throws is 4?

b. What is the probability that the sum of the throws is divisible by 3?

8-5. Tantrum Coin

Bob and George are coins. If you flip Bob any time, the probability that he turns up Heads is ½. If you flip George alone, the probability that he turns up Heads is ½. But if you flip Bob and George together, George has a tantrum and only turns up Tails.

a. Write down the set of possible events if Bob is tossed twice, and then George is tossed twice after Bob.

b. Write down the set of possible events if Bob and George are tossed together, twice.

c. Bob and George's parents prefer Heads to Tails. They award Bob and George one team point for every Head that appears.

    i. What is the probability that Bob and George earn 2 points, if they are each flipped twice, separately?

    ii. What is the probability that Bob and George earn 2 points, if they are flipped twice, together?

    iii. What is the probability that Bob and George earn no points, if they are each flipped twice, separately?

    iv. What is the probability that Bob and George earn no points, if they are flipped twice, together?

8-6. Card Draw

Royce has three cards: an Ace, a King, and a Queen. What is the probability that if Royce draws two cards at random, one of the cards is the Ace?

8-7.   Random Number Generator

If a computer generates random numbers using the four digits 4, 5, 6, and 7, what is the probability that it generates a number less than 5467? (Refer to Problem 1 d).

Note: each digit is used once.

8-8. Independence

Independent events are not affected by other events. Dependent events might be affected by other events. If events have some sort of dependence, the probabilities would change. An example of dependence is the coin in Problem 5 that throws a tantrum. In the following questions, think about whether the events described are independent or dependent.

a. Larry tosses a fair six-sided die four times. The results of the tosses are 2, 6, 4, and 2. What is the probability that the die will come up 5 on Larry's fifth toss?

b. Quarole tosses a fair six-sided die four times. All four times, the die comes up 5. What is the probability that the die will come up 5 on Quarole's fifth toss?

c. Spellman has a plastic die that comes up 5 with a probability of 1/6. But then Spellman leaves the die in his toaster and it melts, so that the 5-face is now together with the 6-face, making an 11-face. The other faces are the same as before. What is the probability that the die comes up 5 after being in the toaster?

d. Harry does not want new shoes. Dudley gets new shoes and brags about how dry and warm his feet feel. What is the probability that Harry now wants new shoes?

e. Harry is allergic to peanuts. Harry sees Dudley eat a whole peanut butter cream pie with great enjoyment. What is the probability that Harry is still allergic to peanuts?

8-9. Favorite Foods

Lori conducted a survey of all fourth-graders at her school. She asked each of them to name their favorite food. Here is her data:

Food	Number of Kids
chow mein	8
cookies	6
cupcakes	3
French fries	4
hamburgers	2
ice cream	6
pizza	13
spaghetti	8
no favorite	10

a. What is the size of Lori's sample space?

b. What is the probability that a random fourth-grader from Lori's school has spaghetti for his or her favorite food?

c. What is the probability that a random fourth-grader from Lori's school has a dessert item for his or her favorite food?

d. Lori wants to find out the probability that the favorite food of a random fourth-grader in the United States is ice cream. What should she do?

8-10. Probability And Statistics

How are statistics estimated using probability? This subject is way beyond this book, but it is so widely used that it is worth mentioning. In many math classes Statistics and Probability are studied at the same time. You can see from Problem 9 how Statistics are used to estimate probabilities and it turns out that probability is used to estimate statistics.

It is almost always impossible to get *all* the data you need for accurate statistics. Suppose, for example, you want to know the average weight of a rabbit in the United States. To answer that question correctly, you'd need to gather the weights of every rabbit in the United States. Getting all that data would be difficult.

Probability can help. Probability can answer the question: how many rabbit-weights do I need in order for me to get a "pretty good" answer.

If you gather the weights of all the rabbits, the probability that your average is correct is 1. If you gather the weights of zero rabbits, the probability that your average weight is correct is zero. Somewhere in between all the rabbits, and zero rabbits is enough rabbits so that the *probability that your average weight is correct* is 0.95, or 95%.

In statistics, we call this a 95% confidence level.

What statisticians do is estimate the number of rabbits needed for 95% using a mathematical model of the rabbit data.

Most of the time, when people are talking about statistics or statistical analysis, they are talking about using statistics and probability together in exactly this way.

If the statistic in question is extremely sensitive or important, we might want to get enough data to reach a 99.9% confidence level. If we just need a general ballpark estimate, we might settle for a 70% confidence level.

Here is a question to think about:

A and B claim to know the average allowance of students at Bellinger High School. A says: "The average allowance is $7.68 per week. I sampled 89 students and my confidence level is 98%." B says: "The average allowance is $6.92 per week. I sampled 145 students." Who has the more accurate average, A or B? Put another way, which number do you believe more, A's number or B's number?

How To Deal With Probability Problems

Learn the vocabulary:

Look for patterns such as:

There might be surprises such as:

9. Bases

10, 1010, 1010, 101010...

A9 + 2 = AB

1+1=10

Bases: Notes and Vocabulary

This chapter is about how to "spell" numbers in different ways. It turns out that the way we "spell" numbers can show off different kinds of patterns and behavior.

You already know, for example, that numbers such as 20, 450, 670, 1000, and 390 are all divisible by 10. That is because in base ten notation, these numbers all end with 0.

The **notation** of a number is how we "spell" it. Here are two notations of the same number: five   and   5.

There are many other ways to notate the number five.

Base ten notation uses the ten Arabic numerals 0, 1, 2, 3, 4, 5, 6, 7, and 9. You already know a lot about base ten notation. You know that the right-most digit is the ones digit, and the digit to the left of the ones digit is the tens digit, and so on.

Base ten notation is also called **decimal notation**, or decimal for short.

Base two notation uses the two Arabic numerals 0 and 1. Base two notation is also called binary notation or binary for short.

You can convert any decimal number to a binary number, and vice versa.

First, to convert whole numbers:

Zero is the same in decimal and binary: 0.
One is the same in decimal and binary: 1.

1 + 1 = two in both decimal and binary. However in binary there is no digit 2. Once you get to two in binary, you run out of digits. When do you run out of digits in decimal?

At number nine. The next whole number, ten, is notated 10 in decimal.

In binary, two is notated 10. One way to think about the meaning of "10" as two is place value. In any base, the right-most digit is the ones digit. In base ten, the next digit is the tens digit. In base two, the next digit is the *twos digit*. The number two is 1 two, 0 ones: 10.

You can write equations in binary:

$$1 + 1 = 10 \quad \text{or} \quad \begin{array}{r} 1 \\ +1 \\ \hline 10 \end{array} \quad \text{see how you "carry 1's" just like in base ten.}$$

By using the rules you already know about addition, you can write the first several binary whole numbers. Fill in the table below.

Number	Binary Sum	Binary Number
two	$\begin{array}{r} 1 \\ +1 \\ \hline 10 \end{array}$	10
three	$\begin{array}{r} 10 \\ +1 \\ \hline \end{array}$	
four	$\begin{array}{r} 10 \\ +10 \\ \hline 100 \end{array}$ or $\begin{array}{r} 11 \\ +1 \\ \hline \end{array}$	100 The next digit in base two is the fours digit!
five	$\begin{array}{r} 11 \\ +10 \\ \hline \end{array}$ or $\begin{array}{r} 100 \\ +1 \\ \hline \end{array}$	
six	$\begin{array}{r} 11 \\ +11 \\ \hline \end{array}$ or $\begin{array}{r} 100 \\ +10 \\ \hline \end{array}$	
seven	$\begin{array}{r} 100 \\ +11 \\ \hline \end{array}$ or	
eight	$\begin{array}{r} 100 \\ +100 \\ \hline \end{array}$ or	
nine	$\begin{array}{r} 110 \\ +11 \\ \hline \end{array}$ or	1001

In the rest of this chapter we'll explore binary arithmetic and other bases. It's important to remember that much of the math you already know—such as long multiplication and long division—applies to numbers in base 2 and other bases.

Here are some binary math questions to get started. Assume all of the numbers on this page are in base 2.

10 – 1 = _____

100 – 10 = _____

10 x 10 = _____

Describe the difference between 1110 and 111.

_____

Describe the difference between 11100 and 111.

_____

How do you write the double of a binary number? For example, how do you write the double of 11010101?

_____

How can you tell if a binary number is odd? _____

How can you tell if a binary number is even? _____

Fill in the table:

Number	Tens Digit in decimal	Ones Digit in decimal	Eights Digit in binary	Fours Digit in binary	Twos Digit in binary	Ones Digit in binary
nine						
twelve						
fifteen						

9-1.   Binary Adding And Subtracting

Try some binary addition and subtraction. Convert the problems to decimal and check your answers.

a.

$$\begin{array}{r} 101 \\ +10 \\ \hline \end{array}$$

b.

$$\begin{array}{r} 1111 \\ +1010 \\ \hline \end{array}$$

c.

$$\begin{array}{r} 1111 \\ -1010 \\ \hline \end{array}$$

d.

$$\begin{array}{r} 1000 \\ -101 \\ \hline \end{array}$$

9-2. Binary Long Multiplication

Try some binary long multiplication. Convert the problems to decimal and check your answers. The first problem is done as an example.

a.

$$
\begin{array}{r}
111 \\
\times\, 11 \\
\hline
111 \\
1110 \\
\hline
10101
\end{array}
$$

b.

$$
\begin{array}{r}
1010 \\
\times\, 10 \\
\hline
\end{array}
$$

c.

$$
\begin{array}{r}
110 \\
\times\, 101 \\
\hline
\end{array}
$$

d.

$$
\begin{array}{r}
1001 \\
\times\, 11 \\
\hline
\end{array}
$$

9-3.   Binary Sequences

Sometimes it is easier to spot a sequence pattern in binary than in decimal, or vice versa. In each of the conversion problems below, ask yourself, what is the pattern of the sequence? Is it easier to see in decimal or in binary?

a. Convert the sequence 1, 2, 4, 8, 16, . . . to binary.

b. Convert the sequence 5, 9, 17, 33, . . . to binary.

c. Convert the binary sequence 1, 100, 111, 1010, 1101, 10000, . . . to decimal. What is the pattern?

9-4. Divisibility In Binary

In the Factors and Primes chapter we discussed divisibility rules for decimal numbers. Do the same rules apply in binary?

a. How can you tell if a binary number is divisible by two?

b. The divisibility rule for three in decimal is: the number is divisible by three if the digits add up to a number divisible by three. Does the same rule apply in binary?

c. How can you tell if a binary number is divisible by four?

9-5.   Place Value In Binary

Binary numbers, like decimal numbers, have place values. A three-digit decimal number such as 457 has a ones digit, a tens digit, and a hundreds digit. The maximum three-digit decimal number is 999, nine hundred and ninety-nine.

a.   What are the three place values in a three-digit binary number?

b.   What is the maximum three-digit binary number? Convert this number to decimal.

c.   What is the maximum four-digit binary number? Convert this number to decimal.

d.   What are the five place values of a five-digit binary number?

e.   What is the maximum five-digit binary number?

f.   (Read). Convert the decimal number 11 to binary using the following method:
      What is the highest power of 2 less than 11 (is it 2, 4, 8, 16, 32)? 8
      Subtract 8 from 11 and you get a remainder of: 3
      The highest power of 2 less than 3 is 2.
      Subtract 2 from 3 and you get a remainder of 1.
      Therefore, 11 = 8 + 2 + 1.
      In binary, eleven = 1000 + 10 + 1 = 1011.

g.   Convert the decimal number 26 to binary using the same method as above:

      What is the highest power of 2 less than 26? _____
      Subtract this power of 2 from 26 and you get a remainder of: _____
      The highest power of 2 less than the remainder is: _____
      Subtract this power of 2 from the remainder and you get a new remainder:
      _____

      Therefore: 26 = _____

9-6.   Binary Long Division

Long division works the same way in binary as in decimal notation. Binary notation can also be used for fractions.

a. (Read) In decimal notation, ½ is 0.5. What is one-half in binary?
    You answer this question the same way you would in decimal. You can use long division.

$$
\begin{array}{r}
0.1 \\
10\overline{)1.0} \\
\underline{1.0} \\
0.0
\end{array}
$$
    The result of long division is 0.1.

b. What is one-quarter in binary?

c. What is three-quarters in binary? (Hint: add up the answers from (a) and (b)).

d. What is one-third in binary?

e. What is one-fifth in binary?

9-7.   Base Three

Base three notation (ternary) uses the three Arabic numerals 0, 1, and 2. Using what you know from the previous work on binary, convert the first ten counting numbers to base three.

Number	Base 3 Sum	Base 3 Number
one	1	1
two	2	2
three	2 +1 — 10	10
four	10 2 +1 or +2	
five	10 +2	
six	12 +1	
seven	11 +10 — 21	21
eight	21 +1 or ?	
nine	22 +1 or ?	
ten	12 +12 or ?	

9-8. Divisibility in Base Three

These questions are about divisibility and division in base three.

a. What divisibility rule is there in base three for numbers that end with the digit 0? Write down a few base three numbers that end in 0 and spot the pattern.

b. How can you tell if a base three number is divisible by two (or even)? (Hint: the role of the number two in base three is similar to the role of the number nine in base ten—two is one less than three, just like nine is one less than ten. Think about the divisibility rule for nine.)

c. What simple fraction is represented by the base-three number 0.1?

d. In any notation system, some fractions are repeating fractions. In decimal, the fraction 1/3 is a repeating fraction, but the fraction 1/5 is not.

   Write the fraction one-half in base three notation. (Use long division). Before you figure it out, do you expect it to repeat or not?

9-9. Base Six Fractions

Base six notation uses the six Arabic numerals 0, 1, 2, 3, 4, and 5.

a. What is the fraction one-sixth in base six? _____

b. What is the fraction one-half in base six? _____

c. What is the fraction one-third in base six? _____

d. Out of one-fourth and one-fifth, which do you expect to be a repeating fraction in base six? Why?

e. The base six number 0.04 corresponds to what simple fraction in lowest terms?

_____

9-10. Hexadecimal

The hexadecimal system is widely used in computer programming. It is base sixteen, and uses the sixteen digits 0, 1, 2, 3, 4, 5, 6, 7, 8, 9, A, B, C, D, E, and F.

a. What is the number six in hexadecimal? _____

b. What is the number twelve in hexadecimal? _____

c. The hexadecimal number 13 is what number in decimal? _____

d. (Read) What is the fraction one-half in hexadecimal? _____

$$\begin{array}{r} 0.8 \\ 2\overline{)1.0} \\ 1.0 \\ \hline 0 \end{array}$$

Notice that in hexadecimal notation:

$$\begin{array}{r} 2 \\ \times\, 0.8 \\ \hline 1.0 \end{array}$$

Because 2 x 8 = sixteen = 10 in hexadecimal.

e. What is the fraction one-sixteenth in hexadecimal? _____

f. What is the fraction one-eighth in hexadecimal? _____
(To check your answer: what is the relationship between one-eighth and one-sixteenth? Compare your answer with the answer to (e).)

g. What is the fraction three-quarters in hexadecimal? _____
(To check your answer: what is the relationship between three-quarters and one-sixteenth?)

h. Think about what fractions repeat and which do not repeat in hexadecimal. What is one-third in hexadecimal?

How To Deal With Base Number Problems

Learn the vocabulary:

Look for patterns such as:

There might be surprises such as:

10. Polyhedra

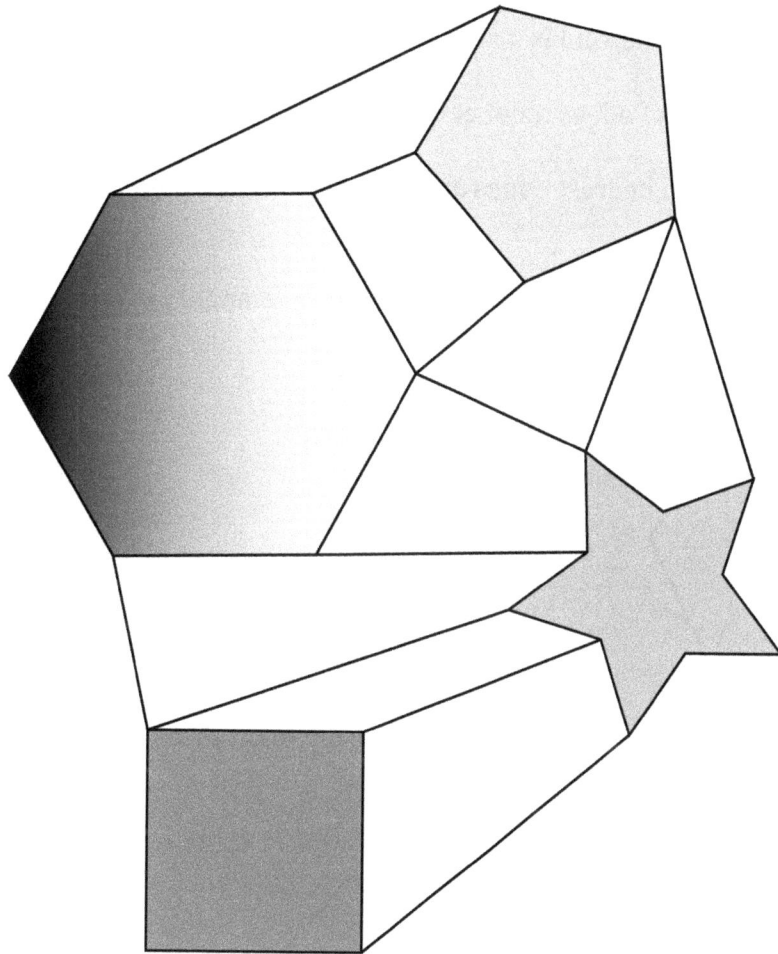

Polyhedra: Notes and Vocabulary
A **polyhedron** is a three-dimensional enclosed surface with flat faces and straight edges. The point where edges meet, a corner, is called a **vertex**.

A familiar example is the cube. Strictly speaking, the word polyhedron refers to the surface of the cube, not the inside of the cube.

Regular Polyhedra
A cube is a member of a special class of polyhedron called the Regular Polyhedra. The Regular Polyhedra have these properties:

- Every face is the same, and is a regular pentagon.

- The same number of edges meet at every vertex.

- The angle between nearest edges at each vertex is the same.

There are only five Regular Polyhedra:
Tetrahedron: four faces, each face is an equilateral triangle.

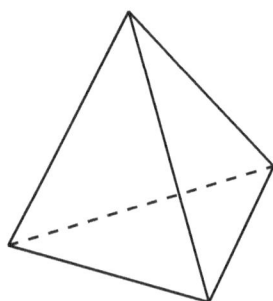

Cube: six faces, each face is a square. Octahedron: eight faces, each face is an equilateral triangle.

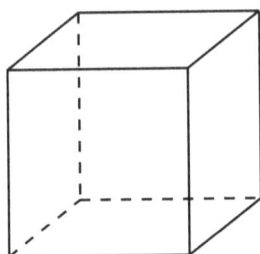

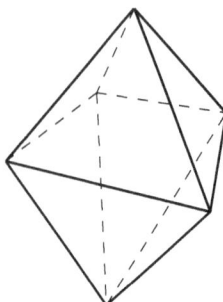

Dodecahedron: twelve faces, each face is a regular pentagon. Icosahedron: twenty faces, each face is an equilateral triangle.

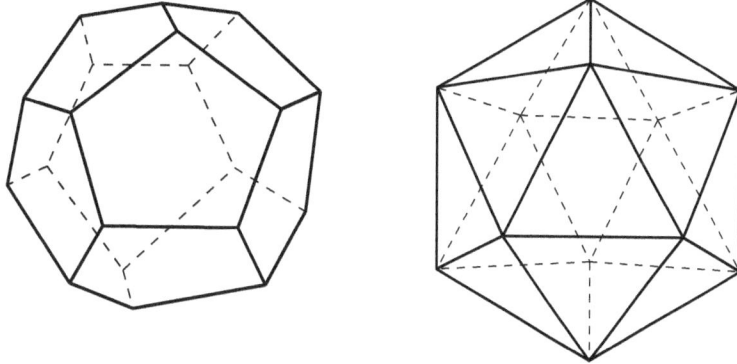

Prisms

A prism is a polyhedron that is similar to a cylinder. The difference between a prism and a cylinder is that if you slice a cylinder, you get a circle, but if you slice a prism, you get a polygon. Here are some examples of prisms.

hexagonal prism

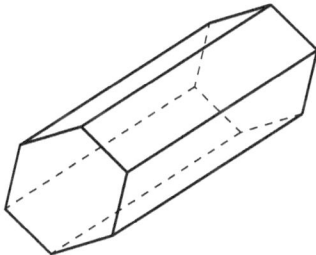

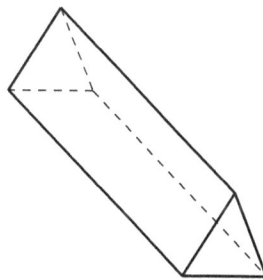

triangular prism

10-1. Cube Wrap

The following diagrams represent pieces of paper cut out along the outer lines. Which pieces of paper could be used to wrap a cube completely without any leftover paper, uncovered areas, overlapping, or tearing?

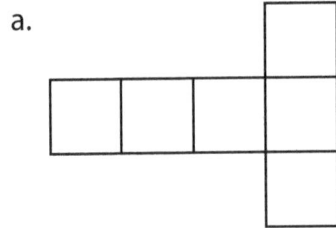

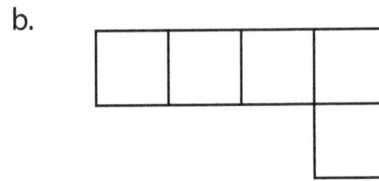

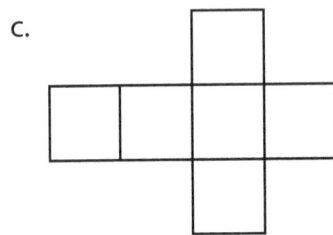

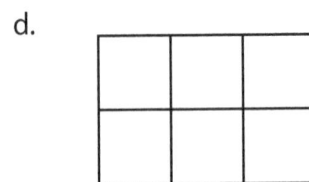

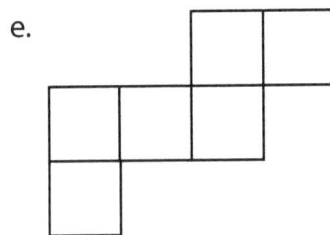

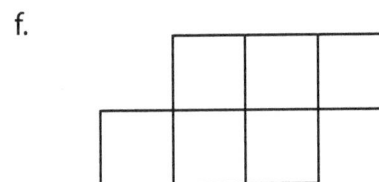

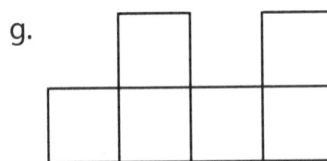

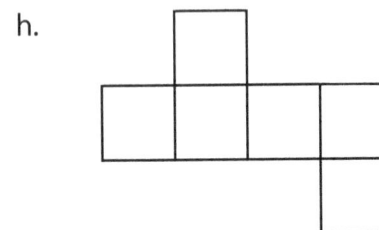

a.

b.

c.

d.

e.

f.

g.

h.

10-2. Face Paint

A large cube is made of little cubes as shown. It is four cubes wide, four cubes long, four cubes deep.

If the large cube is painted all over:

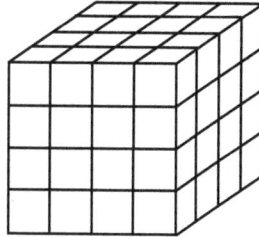

a. How many little cubes have paint on one face only?

b. How many little cubes have paint on two faces only?

c. How many little cubes have paint on three faces only?

d. How many little cubes have no paint at all?

10-3. Three Views Of A Building

Rita uses some small cubes to build a structure. She takes photographs of her structure from three different views.

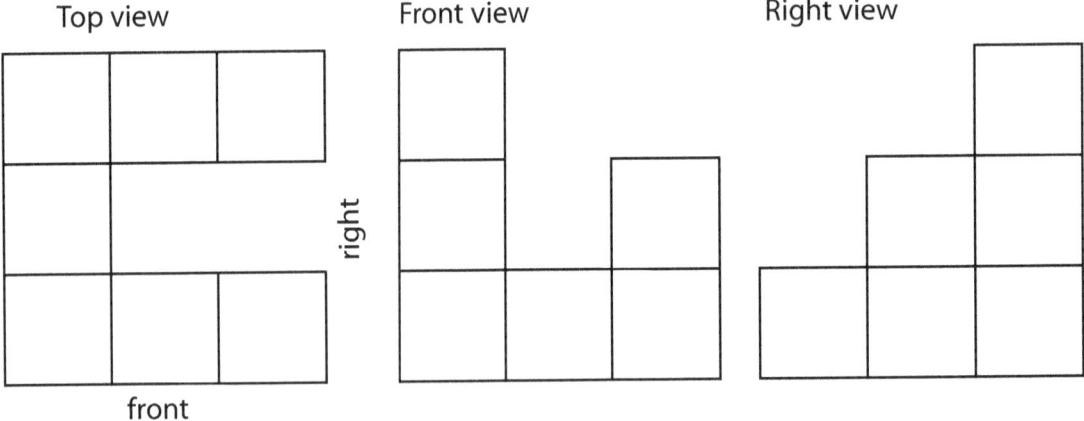

Top view   Front view   Right view

front   right

a. Label the top view with the height of each stack of cubes.

b. How many cubes did she use to build her structure?

10-4.   Cube-O
Randall uses little cubes to build cube-Os.

One-O

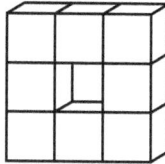

Two-O

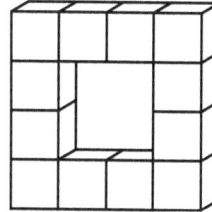

How many cubes does it take to build a Twenty-O?

10-5.   Tetrahedral Dice
Oliver has two tetrahedral dice, each marked 1, 2, 3, 4.

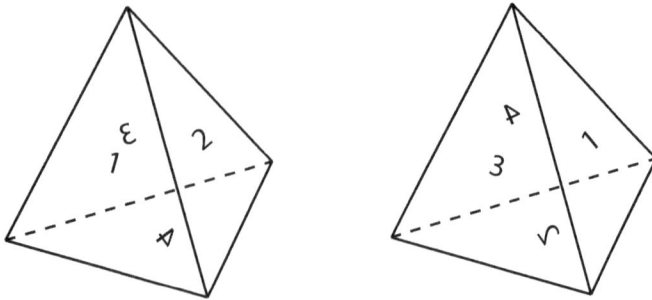

The dice land with one face on the table, and three faces showing.

a.   What is the sample space of one roll of one tetrahedral die?

b.   What is the sample space of one roll of two tetrahedral dice?

c.   What is the probability of rolling a total of 16 in one roll of two tetrahedral dice? (16 would be the sum of all six exposed sides).

d.   What is the total that has the highest probability of occurring in one roll of two tetrahedral dice?

10-6. Cubes Made Out Of Cubes

Pamela has a set of identical wooden cubes one inch long on each side. She'd like to build a sequence of cubes, of side length 1, side length 2, side length 3, side length 4, and side length 5. How many cubes does she need to build them all?

10-7.   Parallelopipeds Out Of Cubes

Louise has 24 identical cubes, each with side length one inch.

a.   How many different solid rectangular polyhedra (parallelopipeds) can she build
     with her cubes? (She must use all 24 cubes each time, but the different
     polyhedra can have different dimensions.)

b.   What is the volume of each polyhedron constructed in part a?

c.   What is the surface area of each polyhedron constructed in part a?

10-8. Euler's Formula

A convex polyhedron is, roughly speaking, a polyhedron that, overall, bulges outward. It does not have any holes or caverns. The five Regular Polyhedra are all convex, and prisms are convex. The Swiss mathematician Leonhard Euler discovered a formula that is true for all convex Polyhedra:

$$V - E + F = 2$$

Where V is the number of vertices, E is the number of edges, and F is the number of faces. Show that Euler's formula is true for all of the Polyhedra below. Count the different numbers of vertices, edges and sides.

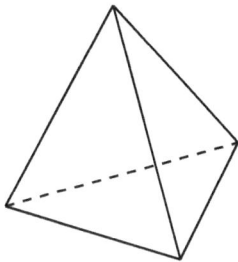

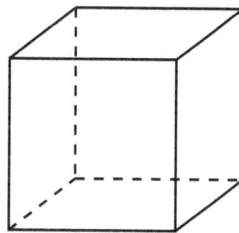

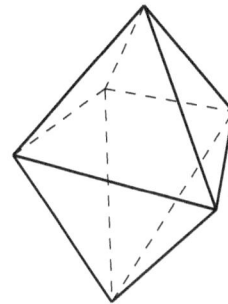

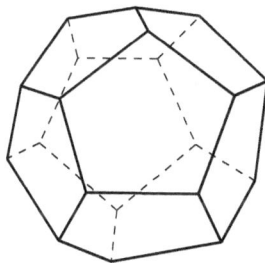

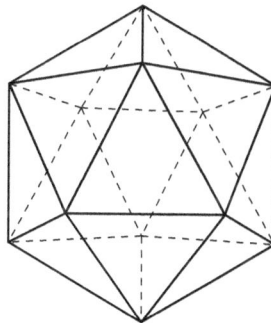

10-9. Octahedron Views

An octahedron is made up of two square pyramids joined together at the base:

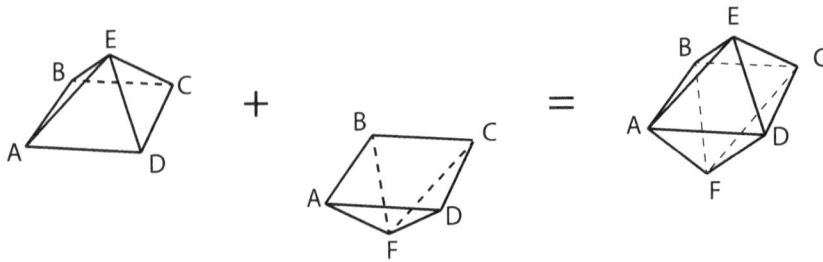

Here are some views of the octahedron from different directions. Label the vertices in these diagrams to match the labeling above.

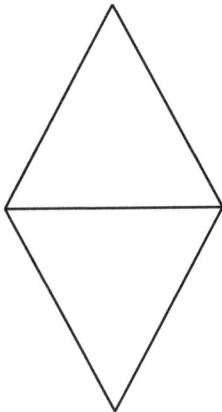

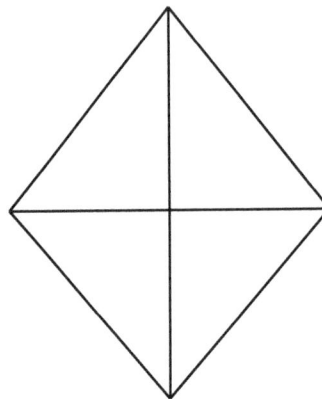

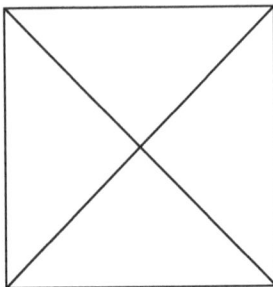

10-10. What Do You Get When You Slice A Cube?

Here are three different ways to slice a cube and get a new polyhedron. In each case, describe the resulting polyhedron. Count how many different polygons form the faces, and name the polygons. Evaluate Euler's formula for each sliced polyhedron.

a. Here is a cube with one corner removed.

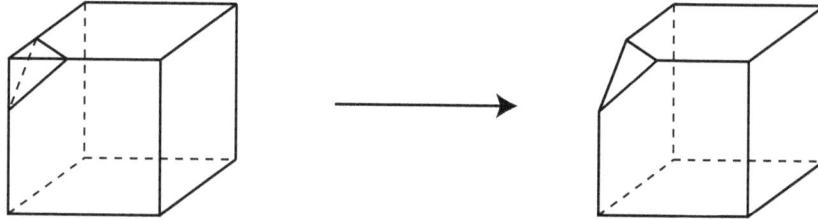

b. Here is a cube sliced in half along the top diagonal.

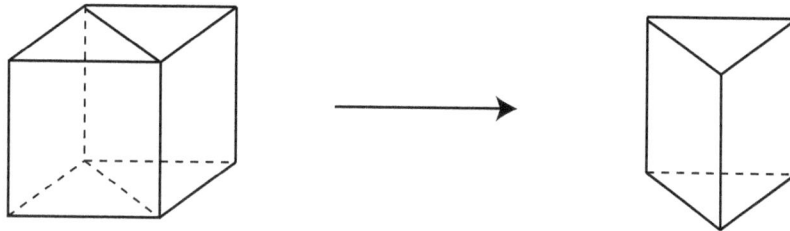

c. Here is a cube sliced at an angle.

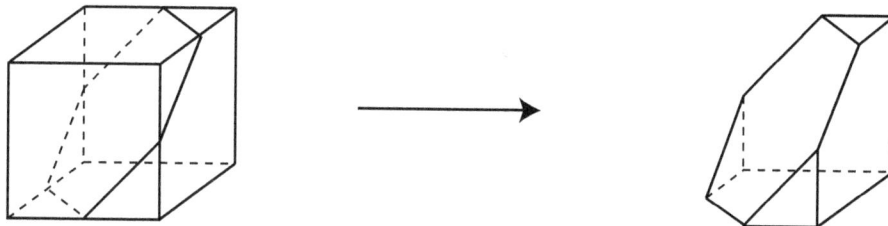

How To Deal With Polyhedra Problems

Learn the vocabulary:

Look for patterns such as:

There might be surprises such as:

Solutions

1. Sequences and Number Patterns

Number	Solution
1-1	a. 6.5, 6.8, 7.1 b. 486, 1458, 4374 c. 27, 35, 44 d. one billion, one trillion, one quadrillion (or one thousand trillion)
1-2	Name three consecutive whole numbers: 5, 6, 7 Name three consecutive whole numbers bigger than a million: two million and one; two million and two; two million and 3 Name three consecutive odd numbers: 11, 13, 15 Name three consecutive multiples of five: 20, 25, 30 Name three consecutive prime numbers: 13, 17, 19 Name three consecutive two-digit primes: 31, 37, 41 Name three consecutive perfect squares: 1, 4, 9
1-3	a. Each term is the sum of the previous two terms. b. 21, 34, 55 c. Fibonacci primes under 100 are: 2, 3, 5, 13, 89. The next one is 233. d. For any three consecutive terms, the square of the middle term has a difference of one from the product of the two outer terms.
1-4	The terms of the sequence $\frac{1}{2}, \frac{1}{4}, \frac{1}{8}, \ldots$ get smaller—closer and closer to zero. The terms of the sequence $\frac{1}{2}, \frac{1}{2}+\frac{1}{4}, \frac{1}{2}+\frac{1}{4}+\frac{1}{8}, \ldots$ get closer and closer to one (they fill up the box).
1-5	a. Variable answer. b. The days of the week repeat after seven days. 100 days from now, the day of the week will be two days from the current day of the week, because 98 is a multiple of seven. c. Yes. 500 days from now, the day of the week will be three days later than the current day, because 497 is a multiple of seven. d. Variable answers, e.g. 777 days from now, the day of the week will be the same as the current day.
1-6	a. 500, b. 1001, c. 2000
1-7	220
1-8	a. 1, 2, 4, 8, 16, ... b. February 9

1-9	a. 9	b. 11
1-10	a. 0, 0, 0, ... b. 1, 2, 4, 8, 16, 9, 2, 4, 8, 16, 9, 2, ... c. 2, 4, 8, 16, 2, ... d. 3, 6, 12, 5, 10, 3, ... e. 4, 8, 16, 9, 2, 4, ... f. 5, 10, 3, 6, 12, 5, ... g. 6, 12, 5, 10, 3, 6, ... h. 7, 14, 7, 14, 7, ... i. 8, 16, 9, 2, 4, 8, ... j. 9, 2, 4, 8, 16, 9, ... The sequences that start with 1, 2, 4, 8, and 9 have similar behavior. The sequences that start with 3, 5, and 6 are similar.	

2. Counting

Number	Solution
2-1	12
2-2	900
2-3	36
2-4	Number of routes: infinite. Number of shortest routes: 6
2-5	114
2-6	a. 20 b. 100 c. 500 d. 2500
2-7	6
2-8	a. 6 b. 24
2-9	28
2-10	12

3. Factors And Primes

Number	Solution
3-1	a. 16 b. 1
3-2	a. 24, with 8 factors. 28 has 6 factors. b. 33, with 4 factors. 31 has 2 factors.
3-3	7, 14, 28
3-4	6
3-5	18
3-6	6
3-7	a. 42, 43 b. 35, 36
3-8	49, 50, 51
3-9	23
3-10	120. None of them is prime, because the sum of the digits is always 18, which means all permutations of 12345 are divisible by 3.

4. Related Numbers

Number	Solution
4-1	a. Drawings vary, one side apple and one side banana and tangerine. b. Yes, as long as the tangerine has some weight. c. Not enough information. d. Not enough information. e. Yes, as long as the banana has some weight. f. Yes, as long as the tangerine has some weight. g. Not enough information. h. Yes, as long as the banana has some weight.
4-2	30
4-3	a. 5 months b. 2 ½ months c. 10 cars per month d. 20 months
4-4	24 cents
4-5	a. 1 fork has 3 ounces of steel. b. 1 spoon has 1 ounce of steel.
4-6	3 miles
4-7	20 days
4-8	a. 3 b. The trick works because you double the input number but subtract it twice, eliminating it. So it doesn't matter what it is. c. No, and no.
4-9	2 hours and 6 minutes
4-10	20 minutes

5. Statistics

Number	Solution
5-1	One out of three data points is an apple: favorite fruit of astronauts. Mode is 10: ages of fifth graders. Mean > 35: average retirement age of police officers. Data between 0 and 130: age of each human being on the planet. Median = 19: number of days books are checked out from library. Mean = 5 million: number of salmon that swim to the mouth of the Skeena. Modes are fedora and porkpie: favorite jazz dancer hats.
5-2	a. 19 b. 65 c. 22

5-3	15
5-4	a. Various, depending on student choices b. mean: 3.6, median: 4, mode: 2 c. There is no obvious best answer here. The data is somewhat skewed to the range from 2 to 4, so the mean might be a best answer. A good thing to do would be to take more measurements and see if a clearer picture develops.
5-5	1, 3, 4, 6, 6
5-6	40
5-7	8
5-8	a. B b. A c. A d. A e. B, because the range of the rainfall is between 0.3 and 0.9, whereas for A, the rainfall has a wider range of 0.1 to 1.0. However, if the month of April is excluded, the range of A is only 0.1 to 0.3. If April is considered an outlier, A has the more even distribution.
5-9	Note: It might help to point out that the expression "compare and contrast" means "look for similarities, and look for differences". The most important difference (in Professor Bear's opinion) is that statement A is limited to 100 dentists who were surveyed. Statement B is far broader, and seems to indicate that 4/5 of all dentists recommend Whizzy-Floss. A smaller but significant difference is that statement A says dentists "preferred" but statement B says dentists "recommend"—two different actions.
5-10	a. mean = 8, median = 8, first mode is 7, second mode is 8 b. Change the first data point from 7 to 2. The mean becomes 7, the median and second mode are unchanged. c. Change the last data point from 10 to 15. The mean becomes 9, the median and first mode are unchanged. d. If there are 100 data points, the last data value would have to go up by 100 to shift the mean up by one unit.

6. Logic

Number	Solution
6-1	A = 1, B = 9, C = 0
6-2	A = 1, B = 0

6-3	

6-4	

What's special about the sums is they are unique. The number 5 can be written as a sum of 1 and 4 or 2 and 3, but 4 can only be a sum of 1 and 3. And so on. |

| 6-5 | Penny: strawberry, Lucy: peach, Max: fig, Carl: raspberry |
| 6-6 | Ciela: 4 brown eggs, Mera: 3 green eggs, Terra: 3 white eggs |
| 6-7 | True. Converse: If the ones digit of a number is 0, 2, 4, 6, or 8, then the number is even. Converse is true if the number is whole.

True. Converse: If a number is not divisible by 4, then it is prime. Converse is false; consider the number 9.

False. Converse: If a number has a ones digit of 1, then the remainder of the number when divided by 5 is 1. Converse is true if the original number is whole. |

6-8	True
	False: Counterexample: 15
	True
	False: Counterexample: googol + 1
	False: Counterexample: 51 is both good and bad.
	False: Counterexample: If Y=6 and X=1, X is less than 3 which is half of Y. Follow up question: for what values of Y is the statement true?
6-9	33
6-10	a. 3 & 5, 11 & 13, 17 & 19, 29 & 31, 41 & 43
	b. There are no other sets of triplet primes. A triplet of primes would be an extension of a pair of twin primes. A pair of twin primes forms the first and third of three consecutive numbers. In any set of three consecutive numbers, there is one multiple of 3 (think about why this is true). In a set of twin primes, the middle number must be the multiple of 3. Any number 2 less than the lower prime or 2 more than the higher prime is a distance of 3 from the middle, and therefore also a multiple of 3. (Diagram these sentences.)
	c. The number between twin primes is divisible by both 3 and 2, so it is divisible by 6.

7. Area And Perimeter

Number	Solution
7-1	a. 24
	b. 4 rectangles: 1x24, 2x12, 3x8, 4x6
	c. 50
	d. 20
7-2	24 cm
7-3	100 square cm
7-4	a. There are three categories of arrangement: short-end to short-end, long-end to long-end, and a combination arrangement where two tiles are horizontal, one vertical.
	b. The possible perimeters are 14 units and 10 units.
	c. There are two possible outer dimensions: 1x6, or 2x3.
7-5	a. 7 square inches b. 16 inches
7-6	This is a way to measure π. Compare how close the numbers in the Circumference/Diameter column come to 3.14159...
7-7	a. 36π
	b. $9\pi^2$ or $(9\pi)x\pi$
	c. Circle
7-8	18 km/h or about 11.3 mph
7-9	Make the argument that the two half-right triangle areas are equivalent to have the area of the outlined rectangle.

7-10	a. 5 sq. units, 25/9 sq. units, 125/81 sq. units, 625/729 sq. units; in sq cm: 36.45, 20.25, 11.25, 6.25 (big square side is 2.7 cm). b. 20 units, 33.333... units, 55.555... units, 92.6 units in cm: 54, 90, 150, 250

8. Probability

Number	Solution
8-1	a. Sample space: red, silver, white, black. Size is 4. b. Sample space: days of the week. Size is 7. c. Sample space: all possible results of three tosses: HHH, HHT, etc. Size is 8. d. Sample space: all permutations of four numbers: 4567, 4576, etc. Size is 24. (A shortcut for counting: there are 6 permutations of three numbers, so 6 possibilities that begin with 4, 6 that being with 5, 6 that begin with 6, 6 that begin with 7.) e. Sample space: radial + gray; radial + aqua, radial + taupe, all-weather + gray, all-weather + aqua, all-weather + taupe. Size is 6.
8-2	10
8-3	18
8-4	a. 1/12 b. 1/3
8-5	a. Bob's results are first two, George's results are second two: HHHH, HTHH, THHH, TTHH, HHTH, HTTH, THTH, TTTH, HHHT, HTHT, THHT, TTHT, HHTT, HTTT, THTT, TTTT b. HHTT, HTTT, THTT, TTTT c. 3/8 d. ¼ e. 1/16 f. ¼
8-6	2/3
8-7	¼
8-8	a. 1/6 b. 1/6 c. zero, since there is no longer a 5-face d. Possibly the probability is now > 0. e. Probability is 1. Nothing has happened to change the allergy.
8-9	a. 60 b. 2/15 c. ¼ d. She needs to get more data.
8-10	B's number.

9. Bases

Number	Solution
9-1	a. 111 (seven) b. 11001 (twenty-five) c. 101 (five) d. 11 (three)
9-2	a. 10101 (twenty-one) b. 10100 (twenty) c. 11110 (thirty) d. 11011 (twenty-seven)
9-3	a. 1, 10, 100, 1000, 10000, ... b. 101, 1001, 10001, 100001, ... c. 1, 4, 7, 10, 13, 16, ... The pattern is to increase by three.
9-4	a. It ends with zero. b. No. c. It ends with zero zero (00).
9-5	a. The three places are: ones, twos, fours. b. The maximum three-digit binary number is 111, which is seven. c. The maximum four-digit binary number is 1111, which is fifteen. d. The five places are: ones, twos, fours, eights, sixteens. e. The maximum five-digit binary number is 11111, which is thirty-one. f. 1011 g. 16, 10, 8, 2, 11010
9-6	a. 0.1 b. 0.01 c. 0.11 d. 0.01010101... Note: leads to infinite series that sums to 1/3 e. 0.001100110011... Note: leads to infinite series that sums to 1/5
9-7	four = 11, five = 12, six = 20, seven = 21, eight = 22, nine = 100, ten = 101
9-8	a. Ends in 0, divisible by three. b. Divisible by two if sum of digits is divisible by two. c. One-third d. Expect one-half to repeat because two shares no factors with three. One-half = 0.111111...
9-9	a. One-sixth = 0.1 b. One-half = 0.3 (one-half is three sixths) c. One-third = 0.2 d. Expect one-fifth to repeat in base six because five shares no factors with six. Expect one-fourth to not repeat because it is the product of one-half with one-half. e. 0.04 = one-ninth

9-10	a. six = 6 b. twelve = C c. 13 = nineteen d. one-half = 0.8 e. one-sixteenth = 0.1 f. one-eighth = 0.2 g. three-quarters = 0.C h. one-third = 0.55555...

10. Polyhedra

Number	Solution
10-1	a. Yes b. No c. Yes d. No e. Yes f. No g. No h. Yes
10-2	a. 24 b. 24 c. 8 d. 8
10-3	a. Top view b. 11
10-4	84
10-5	a. There are 4 elements in the sample space. Three sides at a time are exposed in each roll. The elements are: 123 (adds up to 6), 124 (adds up to 7), 314 (adds up to 8), 234 (adds up to 9) b. The sample space is: 6&6, 6&7, 6&8, 6&9, 7&6, 7&7, 7&8, 7&9, 8&6, 8&7, 8&8, 8&9, 9&6, 9&7, 9&8, 9&9 (sixteen possibilities) c. 3/16 d. 15 has a probability of 1/4
10-6	She needs 1 + 8 + 27 + 64 + 125 = 225 cubes.
10-7	a. 1x1x24, 1x2x12, 1x3x8, 1x4x6, 2x2x6, 2x3x4 (6) b. 24 cubic inches c. 1x1x24: 98, 1x2x12: 76, 1x3x8: 70, 1x4x6: 68, 2x2x6 (56), 2x3x4 (52)

10-8	This problem involves correctly counting the faces, edges, and vertices of each regular polyhedron.
10-9	There are variable answers depending on the point of view.
10-10	a. A polyhedron with 3 square faces, 3 irregular pentagon faces, and 1 triangle face. When evaluating Euler's formula, F=7, V=10, and E=15. b. The resulting polyhedron is a triangular prism, with 2 triangle faces, 2 square faces, and 1 rectangle face (which one is the rectangle face?) When evaluating Euler's formula, F=5, V=6, and E=9. c. The resulting polyhedron has 3 triangular faces, 3 irregular pentagon faces, and one hexagon face. When evaluating Euler's formula, F=7, V=10, and E=15.

Index

www.ingramcontent.com/pod-product-compliance
Lightning Source LLC
Chambersburg PA
CBHW081254040426
42452CB00014B/2501